Selected Titles in This Series

589 **James Damon,** Higher multiplicities and almost free divisors and complete intersections, 1996
588 **Dihua Jiang,** Degree 16 Standard L-function of $GSp(2) \times GSp(2)$, 1996
587 **Stéphane Jaffard and Yves Meyer,** Wavelet methods for pointwise regularity and local oscillations of functions, 1996
586 **Siegfried Echterhoff,** Crossed products with continuous trace, 1996
585 **Gilles Pisier,** The operator Hilbert space *OH*, complex interpolation and tensor norms, 1996
584 **Wayne W. Barrett, Charles R. Johnson, and Raphael Loewy,** The real positive definite completion problem: Cycle completability,, 1996
583 **Jin Nakagawa,** Orders of a quartic field, 1996
582 **Darryl McCollough and Andy Miller,** Symmetric automorphisms of free products, 1996
581 **Martin U. Schmidt,** Integrable systems and Riemann surfaces of infinite genus, 1996
580 **Martin W. Liebeck and Gary M. Seitz,** Reductive subgroups of exceptional algebraic groups, 1996
579 **Samuel Kaplan,** Lebesgue theory in the bidual of $C(X)$, 1996
578 **Ale Jan Homburg,** Global aspects of homoclinic bifurcations of vector fields, 1996
577 **Freddy Dumortier and Robert Roussarie,** Canard cycles and center manifolds, 1996
576 **Grahame Bennett,** Factorizing the classical inequalities, 1996
575 **Dieter Heppel, Idun Reiten, and Sverre O. Smalø,** Tilting in Abelian categories and quasitilted algebras, 1996
574 **Michael Field,** Symmetry breaking for compact Lie groups, 1996
573 **Wayne Aitken,** An arithmetic Riemann-Roch theorem for singular arithmetic surfaces, 1996
572 **Ole H. Hald and Joyce R. McLaughlin,** Inverse nodal problems: Finding the potential from nodal lines, 1996
571 **Henry L. Kurland,** Intersection pairings on Conley indices, 1996
570 **Bernold Fiedler and Jürgen Scheurle,** Discretization of homoclinic orbits, rapid forcing and "invisible" chaos, 1996
569 **Eldar Straume,** Compact connected Lie transformation groups on spheres with low cohomogeneity, I, 1996
568 **Raúl E. Curto and Lawrence A. Fialkow,** Solution of the truncated complex moment problem for flat data, 1996
567 **Ran Levi,** On finite groups and homotopy theory, 1995
566 **Neil Robertson, Paul Seymour, and Robin Thomas,** Excluding infinite clique minors, 1995
565 **Huaxin Lin and N. Christopher Phillips,** Classification of direct limits of even Cuntz-circle algebras, 1995
564 **Wensheng Liu and Héctor J. Sussmann,** Shortest paths for sub-Riemannian metrics on rank-two distributions, 1995
563 **Fritz Gesztesy and Roman Svirsky,** (m)KdV solitons on the background of quasi-periodic finite-gap solutions, 1995
562 **John Lindsay Orr,** Triangular algebras and ideals of nest algebras, 1995
561 **Jane Gilman,** Two-generator discrete subgroups of $PSL(2, R)$, 1995
560 **F. Tomi and A. J. Tromba,** The index theorem for minimal surfaces of higher genus, 1995
559 **Paul S. Muhly and Baruch Solel,** Hilbert modules over operator algebras, 1995
558 **R. Gordon, A. J. Power, and Ross Street,** Coherence for tricategories, 1995
557 **Kenji Matsuki,** Weyl groups and birational transformations among minimal models, 1995

(*Continued in the back of this publication*)

Higher Multiplicities and Almost Free Divisors and Complete Intersections

Number 589

Higher Multiplicities and Almost Free Divisors and Complete Intersections

James Damon

September 1996 • Volume 123 • Number 589 (end of volume) • ISSN 0065-9266

American Mathematical Society
Providence, Rhode Island

1991 *Mathematics Subject Classification.*
Primary 32S30; Secondary 14B05, 58C10.

Library of Congress Cataloging-in-Publication Data

Damon, James, 1945–
Higher multiplicities and almost free divisors and complete intersections / James Damon.
p. cm. – (Memoirs of the American Mathematical Society, ISSN 0065-9266; no. 589)
"September 1996, volume 123, number 589 (end of volume)."
Includes bibliographical references.
ISBN 0-8218-0481-2 (alk. paper)
1. Singularities (Mathematics) 2. Holomorphic mappings. I. Title. II. Series.
QA3.A57 no. 589
[QA614.58]
510 s–dc20
[515′.98] 96-21896
CIP

Memoirs of the American Mathematical Society

This journal is devoted entirely to research in pure and applied mathematics.

Subscription information. The 1996 subscription begins with Number 568 and consists of six mailings, each containing one or more numbers. Subscription prices for 1996 are $391 list, $313 institutional member. A late charge of 10% of the subscription price will be imposed on orders received from nonmembers after January 1 of the subscription year. Subscribers outside the United States and India must pay a postage surcharge of $25; subscribers in India must pay a postage surcharge of $43. Expedited delivery to destinations in North America $30; elsewhere $92. Each number may be ordered separately; *please specify number* when ordering an individual number. For prices and titles of recently released numbers, see the New Publications sections of the *Notices of the American Mathematical Society.*

Back number information. For back issues see the *AMS Catalog of Publications.*

Subscriptions and orders should be addressed to the American Mathematical Society, P. O. Box 5904, Boston, MA 02206-5904. *All orders must be accompanied by payment.* Other correspondence should be addressed to Box 6248, Providence, RI 02940-6248.

Memoirs of the American Mathematical Society is published bimonthly (each volume consisting usually of more than one number) by the American Mathematical Society at 201 Charles Street, Providence, RI 02904-2213. Periodicals postage paid at Providence, Rhode Island. Postmaster: Send address changes to Memoirs, American Mathematical Society, P. O. Box 6248, Providence, RI 02940-6248.

This publication is indexed in *Science Citation Index®, SciSearch®, Research Alert®, CompuMath Citation Index®, Current Contents®/Physical, Chemical & Earth Sciences.*
Printed in the United States of America.
♾ The paper used in this book is acid-free and falls within the guidelines established to ensure permanence and durability.
♻ Printed on recycled paper.

10 9 8 7 6 5 4 3 2 1 00 99 98 97 96

Table of Contents

Abstract

Almost free divisors and complete intersections form a general class of nonisolated hypersurface and complete intersection singularities which simultaneously extend the free divisors introduced by K. Saito and the isolated hypersurface and complete intersection singularities. They also include discriminants of mappings, bifurcation sets, and certain types of arrangements of hyperplanes such as Coxeter arrangements and generic arrangements.

Associated to the singularities of this class is a "singular Milnor fibration" which has the same homotopy properties as the Milnor fibration for isolated singularities. The associated "singular Milnor number" can be computed as the length of a determinantal module using a Bezout-type theorem. This allows us to define and compute higher multiplicities along the lines of Teissier's μ^*-constants.

These are applied to deduce topological properties of singularities in a number of situations including: complements of hyperplane arrangements, various nonisolated complete intersections, nonlinear arrangements of hypersurfaces, functions on discriminants, singularities defined by compositions of functions, and bifurcation sets.

1991 Mathematics subject classification : Primary 32S30

Secondary 14B05, 58C10

Key words and phrases : almost free divisors, singular Milnor fiber, singular Milnor number, algebraic transversality, hyperplane arrangements, Poincaré polynomials, higher multiplicities

To the memory of my parents
James and Dolores Damon

Introduction

Because of the enormous success achieved in understanding the structure of isolated hypersurface singularities, one goal has been to study larger classes of singularities which share important natural features with these singularities. Examples of such classes include isolated complete intersection singularities and Gorenstein surface singularities. The properties of such classes are understood by associating smooth objects such as Milnor fibers or resolutions. For nonisolated singularities, the preceding methods have been applied when the singular behavior can be kept relatively simple (e.g. Siersma, Pellikan et al). However, in many important situations which arise, there are nonisolated singularities with quite complicated singular behavior. It is reasonable to again ask whether there are not natural classes of highly nonisolated singularities which, in fact, share certain nice properties even with isolated hypersurface singularities.

In this paper we shall describe such a class, the class of almost free divisors. In addition to containing isolated hypersurface singularities, this class will also include: discriminants of finitely determined map germs, generalized Zariski examples, bifurcation sets of certain unfoldings of hypersurfaces, certain central arrangements of hyperplanes such as reflection hyperplane arrangements for Coxeter groups or generic arrangements, and nonlinear arrangements of isolated hypersurface singularities.

This class of "almost free divisors" is a natural class of varieties which contains the free divisors and behaves well under pull-backs by algebraically transverse maps and "transverse unions". Furthermore, the intersection of a finite number of such divisors which are in "algebraic general position" leads to a class of almost free complete intersections which includes the isolated complete intersections. In this paper, we shall investigate topological properties for these

---

Partially supported by a grant from the National Science Foundation
Received by the editors November 20, 1993

classes and deduce new results for each of them.

Saito [Sa] intoduced the notion of a free divisor. In joint work [DM], David Mond and this author defined for nonlinear sections f_0: $\mathbb{C}^s,0 \hookrightarrow \mathbb{C}^t,0$ of free divisors $V,0 \subset \mathbb{C}^t,0$ a singular Milnor fibration. Using a result of Lê, it was proven [DM,§4] that the Milnor fiber has the homotopy of a bouquet of spheres of real dimension s-1. Furthermore, the number of such spheres, the "singular Milnor number", could be computed algebraically as a certain codimension [DM,thm 5]. In this paper we will show how these results may be applied to the above examples in unexpected ways.

First, just as Teissier did for isolated singularities [T], it is possible to define higher multiplicities for nonisolated hypersurface singularities $V,0 \subset \mathbb{C}^s,0$ by generic intersection with k-dimensional linear subspaces. In [LêT], Lê and Teissier related the "vanishing Euler characteristics" of generic projections to the properties of the polar curves. Our goal is instead to directly compute the multiplicities as the singular Milnor numbers of the sections. If V is almost free then these multiplicities can be explicitly computed using [DM] as lengths of certain determinantal modules.

A major part of our calculations involve the weighted homogeneous case. In [D4], we gave formulas for the lengths of certain determinantal Cohen-Macaulay modules, the "Macaulay-Bezout numbers", to yield a Bezout-type theorem involving the elementary symmetric functions of the homogeneous degrees in the homogeneous case. For the general weighted homogeneous case, we expressed the weighted Macaulay-Bezout numbers in terms of a universal function τ applied to a "degree matrix" for the module, and deduced algebraic formulas.

In part II, we apply the preceding results to compute the Poincaré polynomials for the complements of arrangements of hyperplanes. We show the fundamental role higher multiplicities play in the topology by giving a general formula for the Poincaré polynomial for hyperplane arrangements in terms of the higher multiplicities. By explicitly computing the higher multiplicities for "almost free arrangements", we are able to give an algebraic formula for Poincaré polynomials for such arrangements in terms of the exponents of the associated free arrangement. These include the free arrangements but also other arrangements such as the "generic arrangements"; in fact, each free arrangement has its own types of generic arrangements. In the case of free arrangements this gives a new proof of the factorization theorem of Terao [To1] [To3] [OT].

These methods have the advantage of being applicable to nonlinear arrangements. There are several possible interpretations of nonlinear arrangements: arrangements of nonlinear hypersurfaces, arrangements of hyperplanes restricted to nonsingular spaces or their Milnor fibers, or a mixture of these. For all of these cases, we can again compute the singular Milnor numbers and higher multiplicities. Using the computations of the Macaulay-Bezout numbers, we compute these numbers for: generic nonlinear arrangements of hypersurfaces of fixed degree based on arbitrary free arrangements A, and generic arrangements of weighted homogeneous hypersurfaces of varying degrees. However, relating them to the Poincaré polynomial for the complement becomes more difficult and is postponed for later consideration.

In part III, We obtain formulas for the singular Milnor numbers and higher multiplicities of *nonisolated* complete intersection singularities which are "almost free". Such complete intersections are represented as algebraically transverse intersections of almost free divisors and include as special cases the ICIS (i.e. isolated complete intersection singularities). Surprisingly, it is the various unions of the almost free divisors whose singular Milnor numbers can be directly computed as lengths of determinantal modules. The singular Milnor number for the almost free complete intersection is given as an alternating sum of singular Milnor numbers for various transverse unions. This becomes very computable because of an observed "principle of linear combination"; namely, each such term can be expressed as a linear combination of fixed terms with varying coefficients. Hence, it is a question of determining by algebraic methods the coefficients of the sum.

In the weighted homogeneous case this can be carried out with Macaulay-Bezout numbers. Using them, we deduce the formulas obtained by Greuel-Hamm [GrH] and Giusti [Gi] for weighted homogeneous ICIS's. We also compute singular Milnor numbers for almost free arrangments on complete intersections, and for the fibers of singular projections of discriminants.

An alternate way to compute singular Milnor numbers is obtained by an analogue of the Lê-Greuel formula for relative Milnor fibers. We again obtain a formula for almost free complete intersections in terms of the length of a determinantal module.

In Part IV, we complete the calculation of "vanishing Euler characteristics" for the pullbacks of almost free divisors and complete intersections. In part I, the computation is given in the case "$n < p$". If "$n > p$" then the pullback need no longer be almost free. Nonetheless, theorem 3 gives a general composition formula for the "vanishing Euler characteristics" in the case "$n \geq p$". This is applied to obtain formulas for the Euler characteristic of nonisolated hypersurface singularities defined via a composition of a germ defining an ICIS with a plane curve singularity; these are the "generalized Zariski examples". Nemethi [N] and Massey-Siersma [MS] obtained formulas involving specific intersection invariants of the curve singularity. We give an alternative general composition formula which involves only the 3 main "Milnor numbers" of the various objects, which can be algebraically computed. Moreover, this composition formula also applies to other situations such as pullbacks of almost free divisors by finite map germs; when applied to modified Zariski examples it provides examples of nonREALizable singular complex cycles, which also occur for bifurcation questions in §11.

In the last section, we consider the bifurcation problems for unfoldings of isolated hypersurface singularities of finite bifurcation codimension and compute the singular Milnor numbers for the bifurcation set. We show that there are universally nonREALizable vanishing cycles for bifurcation problems.

This author would like to thank Mutsuo Oka and Takuo Fukuda for their generous hospitality and financial support during his stay at Tokyo Institute of Technology where some of this work first began. Also, this author would like to acknowledge the support provided by the CNRS and especially the hospitality extended by Prof. Brasselet, the principal organiser for the Congrès Singularités de Lille in 1991, when a number of these results were originally described.

Part I Almost Free Divisors

In this first part we introduce the notion of an almost free divisor as one obtained by the pullback of a free divisor by a germ algebraically transverse to it off the origin. In §1 we recall the principal examples of free divisors and describe the corresponding induced classes of almost free divisors. In §2 we recall certain codimensions which are needed to relate algebraic and geometric transversality. We add for almost free divisors a codimension which provides sufficient numerical conditions for the algebraic properties. In the course of this we compute the "logarithmic tangent space" and deduce fiber square properties for algebraically transverse maps. In §3 we establish several key properties. These include: the "product union" of free divisors is free, the "transverse union" of almost free divisors is almost free, and almost free divisors are preserved under pullbacks by algebraically transverse finite map germs. In §4 we define higher multiplicites for almost free divisors in analogy with Teissier's original definition for isolated hypersurface singularities. Using the results from the preceding sections, we are able to give a simple extension of the results in [DM, thms 5 and 6] to almost free divisors and use it to compute the higher multiplicities in terms of the singular Milnor numbers given by explicit algebraic formulas. We also indicate how these computations can be used to compute the higher multiplicities up to the "free codimension" of a non-almost free divisor.

§1 Free and Almost Free Divisors

We begin by recalling the notion of Free Divisor due to Saito [Sa], giving a number of examples, and extending it to the notion of an Almost Free Divisor. First, we establish some notation.

For a holomorphic germ $f_0 : \mathbb{C}^s,0 \longrightarrow \mathbb{C}^t,0$, the tangent space to the space of germs $C_{(s,t)}$ at f_0 consists of germs of vector fields $\zeta : \mathbb{C}^s,0 \longrightarrow T\mathbb{C}^t$ such that $\pi \circ \zeta = f_0$ (for π: $T\mathbb{C}^t \longrightarrow \mathbb{C}^t$ the projection), and is denoted by $\theta(f_0)$. Thus,

$$\theta(f_0) \cong \mathcal{O}_{\mathbb{C}^s,0}\left\{\frac{\partial}{\partial y_1}, \ldots, \frac{\partial}{\partial y_t}\right\} = \mathcal{O}_{\mathbb{C}^s,0}\left\{\frac{\partial}{\partial y_i}\right\}$$

(we shall denote the R module generated by $\varphi_1,\ldots,\varphi_k$ by $R\{\varphi_1,\ldots,\varphi_k\}$, or $R\{\varphi_i\}$ if k is understood). Also, we let $\theta_s = \theta(id_{\mathbb{C}^s}) \cong \mathcal{O}_{\mathbb{C}^s,0}\left\{\frac{\partial}{\partial x_i}\right\}$ and similarly for θ_t. We also denote the maximal ideal of $\mathcal{O}_{\mathbb{C}^s,0}$ by m_s.

If $M \subset \theta_n$ is an $\mathcal{O}_{\mathbb{C}^n,0}$-module, let $\{\xi_i\}$ denote a set of generators for M. We let $\langle M\rangle_{(x)}$ denote the subspace of $T_x\mathbb{C}^n$ spanned by $\{\xi_{i(x)}\}$. This is only well-defined for x in some sufficiently small neighborhood of 0; however, since all of our statements will only involve results true on a sufficiently small neighborhood, this notation makes sense.

If $(V,0) \subset \mathbb{C}^t,0$ is a germ of a variety, then we consider the module of vector fields tangent to V. Let I(V) denote the ideal of germs vanishing on V. Then (following Saito [Sa]) we let

$$\text{Derlog}(V) = \{\zeta \in \theta_t : \zeta(I(V)) \subseteq I(V)\}.$$

It extends to a sheaf of vector fields tangent to V, $\mathcal{D}erlog(V)$ which is easily seen to be coherent [Sa]. Then, V,0 is called a *Free Divisor* by Saito if Derlog(V) is a free $\mathcal{O}_{\mathbb{C}^t,0}$-module. Its rank is then necessarily t.

Finally it is convenient to define the *logarithmic tangent space to* V at $x \in V$ by

$$T_{\log}(V)_{(x)} \;=\; \langle \mathrm{Derlog}(V)\rangle_{(x)}.$$

This agrees with the usual tangent space at the smooth points of V. As observed in [DM,prop. 3.11], since the elements of Derlog(V) are tangent to the strata of the canonical Whitney stratification of V, if S_i is the canonical Whitney stratum containing x then $T_{\log}(V)_{(x)} \subseteq T_xS_i$.

Examples 1.1:

For any germ $f_0 : \mathbb{C}^s,0 \longrightarrow \mathbb{C}^t,0$ we let $D(f_0)$ denote its discriminant.

a) If F: $\mathbb{C}^{s+q},0 \longrightarrow \mathbb{C}^{s+q},0$ denotes the versal unfolding of an isolated hypersurface singularity, then D(F) is a free divisor by Saito [Sa].

b) More generally. if F denotes the versal unfolding of an isolated complete intersection singularity, then D(F) is a free divisor by Looijenga [L].

c) If B(F) denotes the bifurcation variety of the versal unfolding of an isolated hypersurface singularity, then B(F) is a free divisor by Terao [To2] and Bruce [Br].

d) Isolated curve singularities in $\mathbb{C}^2$ are free divisors by Saito [Sa].

e) Central arrangements of hyperplanes in $\mathbb{C}^n$ which are free divisors when viewed as hypersurfaces are called free arrangements by Terao. He has proven, for example, that the reflection hyperplanes for Coxeter groups are free [To1] [To3], and more generally proves by an inductive process that many other non-reflection arrangements are free.

Next, we enlarge the class of varieties which we consider to the class of almost free divisors. For this, we recall [D1],[D2], and [DM] that f_0: $\mathbb{C}^s,0 \longrightarrow \mathbb{C}^t,0$ is *algebraically transverse to* $V,0 \subset \mathbb{C}^t,0$ *off* 0 if

(1.2)
$$df_0(T_{x_0}\mathbb{C}^s) + T_{\log}(V)_{f(x_0)} \;=\; T_{f(x_0)}\mathbb{C}^t$$

for all x_0 in a punctured neighborhood of 0. We emphasize that (1.2) says

(1.2')
$$df_0(T_{x_0}\mathbb{C}^s) + \mathbb{C}\{\zeta_{1(f(x_0))}, \ldots, \zeta_{m(f(x_0))}\} \;=\; T_{f(x_0)}\mathbb{C}^t .$$

where $\{\zeta_i\}_{i=1}^m$ denotes a set of generators for Derlog(V) (by coherence they also generate the sheaf $\mathcal{Derlog}(V)$ in a neighbourhood of 0). We denote this by writing

$f_0 \stackrel{\circ}{\pitchfork}_{alg} V$ (the "∘" indicates that the transversality is in a punctured neighborhood).

This is related to f_0 being *geometrically transverse to* (V,0) *off* 0, which means that f_0 is transverse to the strata of the canonical Whitney stratification of V at all x_0 in a punctured neighborhood of 0. We denote this by $f_0 \stackrel{\circ}{\pitchfork}_{geom} V$.

By the observation in [DM, prop.3.11], algebraic transversality implies geometric transversality (but not conversely, see lemma 2.10).

Remark: Then, $f_0 \stackrel{\circ}{\pitchfork}_{alg} V$ iff f_0 is finitely $\mathcal{K}_V$-determined.

Here $\mathcal{K}_V$-equivalence is defined via the action of a subgroup of the contact group $\mathcal{K}$

$$\mathcal{K}_V \;=\; \{\Phi \in \mathcal{K}: \Phi(\mathbb{C}^s \times V) \subseteq \mathbb{C}^s \times V\}.$$

This $\mathcal{K}_V$-equivalence captures the isomorphism classes of the germs of varieties $f_0^{-1}(V)$ [D1] [D2].

Although this characterization was stated in [D1] for finite map germs f_0, the proof given there works in general. In fact, by the *graph trick*, every map is equivalent to an embedding by replacing the map f_0 by its graph map $\tilde{f}_0$: $\mathbb{C}^s,0 \longrightarrow \mathbb{C}^s \times \mathbb{C}^t,0$ and replacing V by $\mathbb{C}^s \times V$. If V is free then so is $\mathbb{C}^s \times V$ and f_0 has finite $\mathcal{K}_V$-codimension iff $\tilde{f}_0$ has finite $\mathcal{K}_{\mathbb{C}^s \times V}$-codimension by [D2,§1]. Throughout this paper we will repeatedly reduce proofs to the case of germs of embeddings by the graph trick.

Definition 1.3: A hypersurface $(V',0) \subset \mathbb{C}^s,0$, is an *almost free divisor (based on V)* if there exists a free divisor $(V,0) \subset \mathbb{C}^t,0$ and a germ f_0: $\mathbb{C}^s,0 \longrightarrow \mathbb{C}^t,0$ with $f_0 \stackrel{\circ}{\pitchfork}_{alg} V$ such that $V' = f_0^{-1}(V)$.

Every class of free divisor gives rise to a corresponding class of almost free divisors

Remark : There is also a class of *geometrically almost free divisors* defined as in (1.3) but using geometric rather than algebraic transversality. There are some situations where generically these are the only divisors which are guaranteed to exist. However, at this point there is no known way to compute their vanishing topology and higher multiplicities algebraically.

Examples 1.4:

a) *Discriminants* $D(f_0)$ of finitely ($\mathcal{A}$-) determined germs f_0: $\mathbb{C}^s,0 \longrightarrow \mathbb{C}^t,0$ with $s \geq t$ are almost free divisors by [D2].

b) *Bifurcation Varieties* B(F) of "finite bifurcation-codimension" unfoldings of isolated hypersurface singularities are almost free divisors (see §11).

c) *Isolated Hypersurface Singularities* are almost free divisors because $\{0\} \subset \mathbb{C}$ is a free divisor (Derlog({0}) is freely generated by $z\frac{\partial}{\partial z}$).

d) *Generalized Zariski Examples* are obtained as hypersurface singularities defined by compositions $h \circ f_0$ where h defines an isolated curve singularity in $\mathbb{C}^2,0$, f_0: $\mathbb{C}^s,0 \longrightarrow \mathbb{C}^2,0$ defines an isolated complete intersection singularity, and f_0 is transverse to $h^{-1}(0)$ off 0. Such examples are almost free divisors. Note that each point of $f_0{}^{-1}(0)$ is a singular point with transverse type $h^{-1}(0)$.

e) *An Almost Free Arrangement of Hyperplanes* $A' \subset \mathbb{C}^n$ *based on* a free arrangement $A \subset \mathbb{C}^p$ is defined via $A' = \varphi^{-1}(A)$ for a linear 1-1 map φ: $\mathbb{C}^n \longrightarrow \mathbb{C}^p$ which is transverse to all intersections of subsets of hyperplanes of A(except possibly (0)). Such almost free arrangements are almost free divisors. Then, for example, "generic arrangements" as defined in [OT,5.1] are almost free arrangements based on the Boolean arrangement defined by $\prod z_i = 0$.

An arrangement $A = \cup H_i$ is called *essential* [OT,2.1] if $\cap H_i = (0)$. Any arrangement $A = B \times T$, for an essential arrangement B and $T = \cap H_i$, and $r(A) = \text{codim}(T)$ is called the *rank* of A. More generally given a free arrangement A, we shall say that an arrangement A' is A-generic if $A' = B \times T$ where B is an almost free arrangement based on A.

f) *Nonlinear Almost Free Arrangements* are defined as for arrangements except that $\varphi: \mathbb{C}^n,0 \longrightarrow \mathbb{C}^p,0$ is algebraically transverse to A off 0. For example, we shall see in §3 that the union of a set of isolated hypersurface singularities which are in "algebraic general position off 0" form an almost free nonlinear arrangement; but there are more general examples (see §6).

§2 Algebraic and Geometric Transversality and Codimensions

There is a basic dichotomy between the geometric conditions (e.g. transversality involving V) which we would like to use and the algebraic conditions which we need to apply our results. In this section we will determine numerical bounds on various codimensions which ensure that the algebraic conditions can be established gometrically. For an equation H defining V, we introduce Derlog(H), which is the analogue of Derlog(V) but for H; and we define an H-holonomic codimension $\mathcal{h}(V)$ which is a modified version of the various codimensions considered in [DM,§3]. We derive (corollary 2.16 and proposition 2.19) a sufficient numerical condition, in terms of this codimension, for geometric transversality to imply these stronger algebraic conditions necessary to compute singular Milnor numbers and higher multiplicities for these divisors. At the heart of these arguments are results concerning algebraic transversality to singular varieties (proposition 2.12), especially fiber square arguments (corollaries 2.15).

These problems with various types of transversality don't arise for isolated hypersurface singularities; this is because the numerical condition is always satisfied. However, in general it is a serious problem and is related e.g. to problems such as the existence of topologically stable germs which are not stable [DM,§1].

Various Codimensions

We begin by considering a *good defining equation* for the free divisor $V,0 \subset \mathbb{C}^t,0$. We recall [DM] that this means that V is defined locally by a germ $H: \mathbb{C}^t,0 \longrightarrow \mathbb{C},0$ such that there is a germ of a vector field e with $e(H) = H$. This follows if V is weighted homogeneous, and H can be chosen weighted homogeneous of non-zero weight. However, as observed in [DM], if H defines V, then there is always a good defining equation for $V \times \mathbb{C}$ (namely $H' = e^t H$ where $e = \frac{\partial}{\partial t}$). If V is a free divisor then so is $V \times \mathbb{C}$. Thus, we could replace V by $V \times \mathbb{C}$. A consequence from [DM,lemma 3.3] is that

$$\mathrm{Derlog}(V) = \mathrm{Derlog}(H) \oplus \mathcal{O}_{\mathbb{C}^t,0}\{e\}$$

where $\mathrm{Derlog}(H) = \{\zeta \in \theta_t : \zeta(H) = 0\}$.

Hence, if V is a free divisor then Derlog(H) is a free $\mathcal{O}_{\mathbb{C}^t,0}$-module of rank t−1.

In fact, this is the crucial property, so we also say that H is a *free defining equation for* V if Derlog(H) is a free $\mathcal{O}_{\mathbb{C}^t,0}$-module of rank t−1. Then, any good defining equation is free; however, not always conversely. For example, if h defines an isolated curve singularity $(C,0) \subset \mathbb{C}^2,0$ (which is free by Saito [Sa]) then Derlog(h) is a free $\mathcal{O}_{\mathbb{C}^2,0}$-module generated by $\zeta = \frac{\partial h}{\partial y}\cdot\frac{\partial}{\partial x} - \frac{\partial h}{\partial x}\cdot\frac{\partial}{\partial y}$. However, unless h is weighted homogeneous, this is not a good defining equation.

Just as for Derlog(V), we can define

$$T_{\log}(H)_{(x)} = \langle \mathrm{Derlog}(H) \rangle_{(x)}.$$

Despite the seeming dependence on H we have for good defining equations.

Lemma 2.1 : *For any* $x \in V$, whether $T_{\log}(H)_{(x)} = T_{\log}(V)_{(x)}$ *is independent of the good defining equation* H.

Proof : First, $T_{\log}(H)_{(x)} = T_{\log}(V)_{(x)}$ iff there is an "Euler vector field" e (so that e(H) = H) with $e_{(x)} = 0$. Note "⇐" is immediate. For "⇒", suppose $e_{(x)} \neq 0$. As $e_{(x)} \in \langle \mathrm{Derlog}(H) \rangle_{(x)}$, there is $\zeta \in$ Derlog(H) with $e_{(x)} = \zeta_{(x)}$. Then, $e' = e - \zeta$ is an "Euler vector field" for H and $e'_{(x)} = 0$.

We may assume $e_{(x)} = 0$. If now $H_1 = u \cdot H$ for a unit u, then $e(H_1) = e(u \cdot H) = e(u) \cdot H + u \cdot H = (e(u)+u) \cdot H$. As $e_{(x)} = 0$, $e(u)_{(x)} = 0$ and $u_1 = e(u)+u$ is a unit near x. Then, $e_1 = u_1^{-1} \cdot u \cdot e$ is an "Euler vector field" for H_1 and $e_{1(x)} = 0$. □

If S_i is the canonical Whitney stratum containing x, there are the inclusions

(2.2) $$T_{\log}(H)_{(x)} \subseteq T_{\log}(V)_{(x)} \subseteq T_x S_i.$$

Our codimensions measure when we have equality between these tangent spaces.

For the canonical Whitney stratification $\mathcal{S}$ of V, a stratum $S_i \in \mathcal{S}$ is an *H-holonomic stratum* if for all $x \in S_i$,

(2.3) $$T_xS_i = T_{\log}(H)_{(x)}.$$

It is a *holonomic stratum* if (2.3) holds instead for $T_{\log}(V)_{(x)}$; and it is *weighted homogeneous stratum* if near each $x \in S_i$ there are local coordinates about x so that (V, x) is weighted homogeneous near x. We observe that if V is weighted homogeneous near x, then there is an H and an "Euler vector field" e so that e(H) = H and $e_{(x)} = 0$. Hence,

(2.4) $$T_{\log}(H)_{(x)} = T_{\log}(V)_{(x)}.$$

Now we can define the various codimensions.

$$\mathcal{h}(V) \overset{\text{def}}{=} \max\{k: \text{all strata } W_i \text{ of codimension} < k \text{ are H-holonomic}\}$$

i.e., $\mathcal{h}(V)$ is the codimension of the largest stratum which is not H-holonomic. We similarly define *holonomic codimension* $\mathcal{hn}(V)$ and the *weighted homogeneous codimension* $\mathcal{wh}(V)$. By (2.3) and (2.4) we have

(2.5) $$\min\{\mathcal{hn}(V), \mathcal{wh}(V)\} \leq \mathcal{h}(V) \leq \mathcal{hn}(V)$$

If all strata are holonomic then following Saito [Sa] (see also [BR]) we say that V is *holonomic* and write $\mathcal{hn}(V) = \infty$ (i.e. the empty stratum has infinite codimension). We analogously have the notion of being H-*holonomic.*

Remark : In what follows we shall use a free defining equation for which the holonomic codimension is smallest. If this is a good defining equation then by lemma 2.1, it is independent of the equation. *In all cases we consider except one,* namely isolated curve singularities, we will use a good defining equation for V. However, isolated curve singularities are H-holonomic for the original defining equation h, so in this case we use this free defining equation.

Remark : We have used script letters $\mathcal{hn}(V)$ and $\mathcal{wh}(V)$ to emphasize for readers that these codimensions will be "off by one" from those used in [DM]. This change is being made to correct a discrepancy in wording; namely, if a property holds in the complement of $\{0\} \subset \mathbb{C}^q$ then the codimension for the property should be q, but it would not be with the original definitions.

For an almost free divisor V',0 defined from V,0, we extend this to a codimension $\tilde{h}(V')$.

(2.6) $$\tilde{h}(V') \stackrel{\text{def}}{=} \min\{hn(V'), h(V)\}$$

We will show that provided we remain below $\tilde{h}(V')$, the algebraic conditions can be established geometrically. We let $\{\zeta_1, \ldots, \zeta_{t-1}\}$ denote the generators for Derlog(H) and let $\mathbb{C}^t$ have coordinates $\{y_1, \ldots, y_t\}$. Then we define what will be an algebraic formula for the singular Milnor number for the nonlinear section $f_0 : \mathbb{C}^s,0 \to \mathbb{C}^t,0$ of V.

(2.7) $$\nu_V(f_0) = \dim_{\mathbb{C}} \mathcal{O}_{\mathbb{C}^s,0}\left\{\frac{\partial}{\partial y_i}\right\} / \left(\mathcal{O}_{\mathbb{C}^s,0}\left\{\frac{\partial f_0}{\partial x_i}\right\} + \mathcal{O}_{\mathbb{C}^s,0}\left\{\zeta_1 \circ f_0, \ldots, \zeta_{t-1} \circ f_0\right\}\right)$$

Remark: It is not always true that if $f_0 \mathring{\pitchfork}_{alg} V$ then $\nu_V(f_0) < \infty$. If we define

$$\mathcal{K}_H = \{\Phi \in \mathcal{K} : H \circ \Phi = H\}$$

then this group measures the $\mathcal{K}$-equivalence of germs by an equivalence preserving all of the level sets of H and $\nu_V(f_0)$ is the $\mathcal{K}_{H,e}$-codimension of f_0. There is an analogous characterization of finite determinacy for $\mathcal{K}_H$ [DM,§3]:

(2.8): f_0 is finitely $\mathcal{K}_H$-determined iff the analogue of (1.2) is valid for x_0 in a punctured neighborhood of 0, where we use $\{\zeta_1, \ldots, \zeta_{t-1}\}$.

Definition 2.9 : With the above notation, we will say that $V' = f_0^{-1}(V)$ is *finitely defined* if $\nu_V(f_0)$ (= $\mathcal{K}_{H,e}$-codim(f_0)) $< \infty$.

Remark : The algebraic transversality conditions on f_0 ensure that it has finite $\mathcal{K}_{V,e}$-codimension but not necessarily finite $\mathcal{K}_{H,e}$-codimension, since the first inclusion in (2.2) can be proper. We shall give in corollary 2.16 a sufficient numerical condition for geometric transversality to imply this.

Lemma 2.10 : *Suppose* $V,0 \subset \mathbb{C}^t,0$ *a free divisor and* $f_0 : \mathbb{C}^s,0 \to \mathbb{C}^t,0$.

i) $f_0 \mathring{\pitchfork}_{alg} V \Rightarrow V' = f_0^{-1}(V)$ *is finitely defined if either*

a) *both* f_0 *and* V *are weighted homogeneous for the same weights, or*

b) $s \leq h(V)$.

ii) *If* $s \leq hn(V)$ *then algebraic and geometric transversality off 0 coincide.*

iii) *If* $s \leq h(V)$ *then* $f_0 \mathring{\pitchfork}_{geom} V \Rightarrow V' = f_0^{-1}(V)$ *is finitely defined.*

Proof : For a) we use prop. 3.23 of [DM] to conclude the two codimensions are equal. For b), since $f_0 \mathring{\pitchfork}_{alg} V$ implies $f_0 \mathring{\pitchfork}_{geom} V$, the condition $s \leq h(V)$ implies that in a punctured neighborhood of 0, $f_0(\mathbb{C}^s)$ only intersects H–holonomic strata of V and intersects them transversely. Hence, (2.5) implies that (2.8) holds at points $x_0 \neq 0$ where $f_0(x_0) \in V$. Then, to conclude that it could not fail at other points we use an observation of David Mond. Consider a curve $\gamma(t)$ of points through 0 where (2.8) fails and $f_0(\gamma(t)) \notin V$, all $t \neq 0$. For $y \notin V$ the level surface of H through y is a smooth hypersurface with tangent space = $T_{log}(H)_{(y)}$. Thus, transversality can only fail by $df_0(\gamma'(t)) \in T_{log}(H)_{(f_0(\gamma(t)))}$. This implies $H \circ f_0(\gamma(t))$ is constant = H(0) = 0, a contradiction.
Then, ii) follows from prop.3.11 of [DM]; and iii) follows from i) and ii). □

Examples 2.11 H–Holonomic codimensions of free divisors :

a) and b) If $F: \mathbb{C}^{s+q},0 \rightarrow \mathbb{C}^{s+q},0$ is a versal unfolding of an ICIS then $h(D(F)) \geq \text{codim}(NS_{\mathcal{K}}(F))$ where $NS_{\mathcal{K}}(F) = \{\ \mathcal{K}\text{-nonsimple points } y \in D(F)\}$. We apply (5.25) using $hn(D(F)) = \text{codim}(NS_{\mathcal{K}}(F))$ while $wh(D(F)) \geq \text{codim}(NS_{\mathcal{K}}(F))$ by Mather's classification [M–VI].

c) For F the versal unfolding of an isolated hypersurface singularity, $h(B(F)) \geq \text{codim}(NS_{\mathcal{R}}(F))$ where $NS_{\mathcal{R}}(F) = \{\ \mathcal{R}\text{-nonsimple points } u \in B(F)\}$. Again $hn(B(F)) = \text{codim}(NS_{\mathcal{R}}(F))$ and by the classification [A] $wh(B(F)) \geq$

codim($NS_{\mathcal{R}}(F)$).

d) Isolated curve singularities $C \subset \mathbb{C}^2$ are H-holonomic since if C is defined by h then $\zeta = \frac{\partial h}{\partial y}\cdot\frac{\partial}{\partial x} - \frac{\partial h}{\partial x}\cdot\frac{\partial}{\partial y}$ generates Derlog(h) and is nonzero on $C\backslash\{0\}$.

e) All central arrangements are holonomic by [To1]; here the idea is simply that, off 0, A is locally affinely equivalent to $W \times A'$ where the ranks satisfy $r(A') < r(A)$. Hence , it follows by coherence of *Derlog(A)* and induction on rank. Also, $wh(A) = \infty$ since central arrangements are homogeneous, so they are all H-holonomic so $h(A) = \infty$.

Properties of Algebraic Transversality to Singular Varieties

Next, for almost free divisors we determine both the logarithmic tangent spaces and deduce that certain standard transversality arguments can be applied to singular varieties. These will then allow us to bound $\tilde{h}(V')$ for almost free divisors.

We will consider $V, 0 \subset \mathbb{C}^t, 0$ (not necessarily free) and $f_0: \mathbb{C}^s, 0 \longrightarrow \mathbb{C}^t, 0$ with $f_0 \ \mathring{\pitchfork}_{alg}\ V$ so that $V' = f_0^{-1}(V)$. We define

$$tf_0^{-1}(\mathrm{Derlog}(V)) = \{\xi \in \theta_s : tf_0(\xi) \ (\overset{def}{=} df_0(\xi)) = \eta \circ f_0 \text{ for } \eta \in \mathrm{Derlog}(V)\}.$$

Proposition 2.12: *Then,* $tf_0^{-1}(\mathrm{Derlog}(V)) \subseteq \mathrm{Derlog}(V')$ *differ by finite codimension. Hence there is a* $k > 0$ *so that*

$$m_s^k \cdot \mathrm{Derlog}(V') \subseteq tf_0^{-1}(\mathrm{Derlog}(V)); \qquad \textit{and thus,}$$

$$(2.13) \qquad T_{log}(V')_{(x)} = df_0(x)^{-1}(T_{log}(V)_{(f(x))})$$

Before proving this proposition, we deduce two important consequences.

Fiber squares for Algebraic Transversality:

We consider the diagram of mappings (2.14) with f_0 a finite map germ, each of i, i′, and i″ denoting inclusions, and both $V' = g_0^{-1}(V)$ and $V'' = f_0^{-1}(V')$.

fig. 2.14

$$\begin{array}{ccccc} & f_0 & & g_0 & \\ \mathbb{C}^n,0 & \longrightarrow & \mathbb{C}^p,0 & \longrightarrow & \mathbb{C}^m,0 \\ \uparrow i'' & & \uparrow i' & & \uparrow i \\ V'',0 & \longrightarrow & V',0 & \longrightarrow & V,0 \end{array}$$

Corollary 2.15: *Given the mappings in fig. 2.14, with indicated properties, we also suppose* $g_0 \mathring{\pitchfork}_{alg} V$. *Then* $f_0 \mathring{\pitchfork}_{alg} V'$ iff $g_0 \circ f_0 \mathring{\pitchfork}_{alg} V$.

Furthermore, by bringing together this corollary with lemma 2.10 we obtain

Corollary 2.16: *Suppose the mappings in fig. 2.14, satisfy* $g_0 \mathring{\pitchfork}_{alg} V$, *but with only* $f_0 \mathring{\pitchfork}_{geom} V'$. *If* $n \le \tilde{h}(V')$ *then* V'' *is finitely defined.*

Proof (of Corollary 2.15): As f_0 is a finite map germ, $f_0^{-1}(0) = 0$. If $x \neq 0$ then the corresponding maps of tangent spaces is given in fig. 2.17 where the

fig. 2.17

$$\begin{array}{ccccc} & df_0 & & dg_0 & \\ T_x\mathbb{C}^n,0 & \longrightarrow & T_{(f_0(x))}\mathbb{C}^p,0 & \longrightarrow & T_{(g_0 \circ f_0(x))}\mathbb{C}^m,0 \\ & & \uparrow i' & & \uparrow i \\ & & T_{log}(V')_{(f_0(x))} & \longrightarrow & T_{log}(V)_{(g_0 \circ f_0(x))} \end{array}$$

square becomes a fiber square of vector spaces by proposition 2.14. Hence, corollary 2.15 follows by standard linear algebra. □

Proof (of Corollary 2.16): As $n \le \tilde{h}(V') \le hn(V')$, by ii) of lemma 2.10,

$f_0 \,\mathring{\pitchfork}_{geom}\, V'$ implies $f_0 \,\mathring{\pitchfork}_{alg}\, V'$. Then, proposition 2.12 implies $g_0 \circ f_0 \,\mathring{\pitchfork}_{alg}\, V$. Lastly, as $n \leq \tilde{\hbar}(V') \leq \hbar(V)$, by ib) of lemma 2.10, V' is finitely defined. □

Proof (of proposition 2.12):

Step 1: We reduce to a local embedding. We may apply the graph trick and assume that $f_0: \mathbb{C}^s,0 \longrightarrow \mathbb{C}^t,0$ is an embedding, which, by a local change of coordinates, we take to be the standard embedding. Also, we may pick a neighborhood $U \subset \mathbb{C}^t$ of 0 on which a set of generators $\{\zeta_i\}$ for Derlog(V) are defined; furthermore, by the coherence of $\mathcal{D}erlog(V)$ we may assume that $\{\zeta_i\}$ generate Derlog(V,x) for all $x \in U$. Moreover, by $f_0 \,\mathring{\pitchfork}_{alg}\, V$, there is a neighborhood of 0 in $\mathbb{C}^s$, which we may assume is $U \cap \mathbb{C}^s$, so that (1.2) holds for all $x_0 \in U \cap \mathbb{C}^s \backslash \{0\}$.

Step 2: For any $x_0 \in U \cap \mathbb{C}^s \backslash \{0\}$,

$$(V, f_0(x_0)) \xrightarrow{\sim} (V' \times \mathbb{C}^{t-s}, (x_0, 0))$$

By the algebraic transversality to V at x_0, there is a minimal subset $\{\zeta_{i_1}, \dots, \zeta_{i_{t-s}}\} \subseteq$ Derlog(V) so that

$$\textbf{(2.18)} \qquad df_0(T_{x_0}\mathbb{C}^s) + \mathbb{C}\{\zeta_{i_1(f(x_0))}, \dots, \zeta_{i_{t-s}(f(x_0))}\} = T_{f(x_0)}\mathbb{C}^t.$$

As the $\{\zeta_{i_1(f(x_0))}, \dots, \zeta_{i_{t-s}(f(x_0))}\}$ are linearly independent and transverse to $df_0(T_{x_0}\mathbb{C}^s)$ by (2.18), by integrating $\{\zeta_{i_1}, \dots, \zeta_{i_{t-s}}\}$ we obtain $(V, f_0(x_0)) \xrightarrow{\sim} V' \times \mathbb{C}^{t-s}, (x_0, 0)$.

Step 3: For the proof proper, we first claim that $tf_0^{-1}(\text{Derlog}(V)) \subseteq \text{Derlog}(V')$. Let $I = I(V)$. If $\xi \in tf_0^{-1}(\text{Derlog}(V))$ then $df_0(\xi) = \eta \circ f_0$ for $\eta \in$ Derlog(V). Then, $\eta(I) \subseteq I$ and so if $g \in I$, $\xi(f_0^*(g)) = df_0(\xi)(g) = \eta(g) \circ f_0 \in f_0^*(I)$, i.e. $\xi(f_0^*(I)) \subseteq f_0^*(I)$.

By applying step 2 to x in the punctured neighborhood $U \cap \mathbb{C}^s \backslash \{0\}$ of 0,

$I(V',x) = f_0^*(I)_x$. Thus, if $g \in I(V')$, then as a germ $g_{(x)} \in f_0^*(I)_x$. Hence, $\xi(g_{(x)}) \in f_0^*(I)_x$, or $\xi(g)$ vanishes on V' in a punctured neighborhood of 0 so by continuity $\xi(g) \in I(V')$. Thus, $\xi \in \mathrm{Derlog}(V')$.

Step 4 : To prove the finite codimension condition, we sheafify these modules to obtain $tf_0^{-1}(\mathcal{D}erlog(V)) \subseteq \mathcal{D}erlog(V')$. However, by step 2, if $x_0 \neq 0$, then $(V, f_0(x_0)) \xrightarrow{\sim} (V' \times \mathbb{C}^{t-s}, (x_0, 0))$; thus, $tf_0^{-1}(\mathcal{D}erlog(V))_{(x_0)} = \mathcal{D}erlog(V')_{(x_0)}$ or $\mathrm{supp}(\mathcal{D}erlog(V')/tf_0^{-1}(\mathcal{D}erlog(V))) = \{0\}$. Then the finite codimension follows from the Nullstellensatz for coherent analytic sheaves (see e.g. [Tg,chap.2,§7]). Lastly, by Nakayama's lemma there is a $k > 0$ so that $m_s{}^k.\mathrm{Derlog}(V') \subseteq tf_0^{-1}(\mathrm{Derlog}(V))$. Since $\mathrm{supp}(\mathcal{O}_{\mathbb{C}^s,0}/m_s{}^k) = \{0\}$, the last statement follows by evaluating for x in a punctured neighborhood of 0

$$<m_s{}^k.\mathrm{Derlog}(V')>_{(x)} \subseteq <tf_0^{-1}(\mathrm{Derlog}(V))>_{(x)} \subseteq <\mathrm{Derlog}(V')>_{(x)} \quad \square$$

Codimensions for Almost Free Divisors

Lastly, we apply the preceding to obtain lower bounds for $\tilde{h}(V')$ for an almost free divisor V' so we can apply iii) of lemma 2.10. We again consider $V, 0 \subset \mathbb{C}^t,0$ (not necessarily free), $f_0: \mathbb{C}^s,0 \longrightarrow \mathbb{C}^t,0$ with $f_0 \ \mathring{\pitchfork}_{alg}\ V$, and $V' = f_0^{-1}(V)$. We let $gm(V)$ denote the codimension of the canonical Whitney stratum containing 0; this is in one sense the intrinsic "geometric codimension" of (V,0).

Proposition 2.19: *For* $V', 0 \subset \mathbb{C}^s,0$ *defined above ,*

i) $hn(V') \geq hn(V)$ *if either* $hn(V) \leq gm(V)$ *or* $s > gm(V)$;

and in the latter case, if V *is holonomic so is* V',

while $hn(V') \geq s$ *if* $s \leq gm(V)$ *and* V *is holonomic*

ii) *Hence, when* V *is a free divisor, so* V' *is almost free,*

$$\tilde{h}(V') = \begin{cases} h(V) & \text{if } hn(V) \leq gm(V) \text{ or } s > gm(V) \\ \min\{s, h(V)\} & \text{if } s \leq gm(V) \text{ and V is holonomic.} \end{cases}$$

and

This yields the following

Examples 2.20: Holonomic codimensions of almost free divisors.

As a consequence of proposition 2.19, we see that the generalized Zariski examples ($s > 2$), almost free arrangements, and isolated hypersurface singularities are holonomic and so $\tilde{h} = \infty$. Nonlinear arrangements based on A are holonomic if $s > r(A) = gm(A)$, or $hn = s$ if $s \leq r(A)$; in either case $\tilde{h} \geq s$. Thus, pull backs of these almost free divisors by finite germs geometrically transverse off 0 will be almost free.

For finitely ($\mathcal{A}$-) determined germs $f_0: \mathbb{C}^s,0 \to \mathbb{C}^t,0$, $\tilde{h}(D(f_0)) \geq \text{codim}(NS_{\mathcal{K}}(f_0))$. An analogous formula in bifurcation holds, $\tilde{h}(B(F_0)) \geq \text{codim}(NS_{\mathcal{R}}(F))$. See example 2.11.

To prove proposition 2.19, we first collect together in the following lemma several elementary observations about holonomic codimension.

Lemma 2.21: *Given* $V,0 \subset \mathbb{C}^t,0$, *then*

i) *if* $(V',x') \xrightarrow{\sim} (V \times \mathbb{C}^q, (x, 0))$ *then* $hn(V',x') = hn(V,x)$; *and*

ii) *if* $k \leq gm(V)$, *then*

$hn(V) \geq k$ *iff* $hn(V,x') \geq k$ *for all* x' *in a punctured neighborhood of* 0.

Proof: For i) we need only observe that the canonical Whitney stratification for $V \times \mathbb{C}^q$ is the product of that for V with $\mathbb{C}^q$, and S_i is holonomic iff $S_i \times \mathbb{C}^q$ is.

For ii) the equivalence follows because whether $hn(V) \geq k$ depends on whether the canonical strata S_i of codimension $i < k$ are holonomic. However, $i < k \leq gm(V)$ implies $0 \notin S_i$; hence, whether S_i is holonomic is determined by the local behavior at each $x \in S_i$ since the construction of the canonical stratification is natural for restriction to open sets. □

Proof (of proposition 2.19):

By step 1 of the proof of proposition 2.12, we assume there is a neighborhood U of $0 \in \mathbb{C}^t$ so that: i) $f_0: U \cap \mathbb{C}^s \to \mathbb{C}^t$ is the standard embedding; ii) generators $\{\zeta_i\}$ for Derlog(V) are defined on U so that $\{\zeta_i\}$ generate Derlog(V,x) for all $x \in U$; and iii) algebraic transversality of f_0 to V holds on $U \cap \mathbb{C}^s \backslash \{0\}$. Hence, by integrating $\{\zeta_{i_1}, \dots, \zeta_{i_{t-s}}\}$ we obtain $V, f_0(x_0) \xrightarrow{\sim} V' \times \mathbb{C}^{t-s}, (x_0, 0)$.

By i) of lemma 2.21 together with step 2 of the proof of (2.12), we obtain $hn(V', x_0) = hn(V, f_0(x_0))$. Thus,

(2.22) $\quad hn(V) \geq k \quad \Rightarrow \quad hn(V') \geq k \quad$ for all $k \leq gm(V)$.

By transversality off 0, the pull-back by f_0 of the canonical stratification will be the canonical stratification off 0; thus, $gm(V') \geq gm(V)$. Thus, if $hn(V) \leq gm(V)$, then $hn(V') \geq hn(V)$.

If $s > gm(V)$ and V is holonomic,then we claim V′ is holonomic. First, let V_0 be the canonical stratum of V containing 0; it is holonomic by assumption. As $T_0V_0 = T_{\log}(V)_{(0)}$, V is analytically a product along V_0, say $V, 0 \xrightarrow{\sim} V_0 \times W, 0$. If f_0 is transverse to V_0 at 0, then $V', 0 \xrightarrow{\sim} V_0' \times W, 0$, so $hn(V') = hn(W) = hn(V)$ by i) of lemma 2.21. Otherwise, $\{0\}$ is an isolated singular point of $f_0^{-1}(V_0)$, and hence, a separate stratum. Morover, $f_0^{-1}(V_0)\backslash\{0\}$ is holonomic by the above remarks; thus, $hn(V') = \infty$.

Lastly, suppose $s \leq gm(V)$ and V is holonomic. Then, for V_0 the canonical stratum containing 0, $f_0^{-1}(V_0) = \{0\}$; while (2.22) shows all other strata are holonomic so $hn(V) \geq s$.

§3 Properties of Almost Free Divisors

We derive several additional properties of almost free divisors. We introduce the notion of a "product union" and prove that the product union of free divisors is free; using this we futher prove that the "transverse union" of almost free divisors is almost free; and thirdly, we prove that almost free divisors are preserved under the pullbacks by finite map germs algebraically transverse off 0.

As already remarked, if $V,0 \subset \mathbb{C}^t,0$ is free then $V \times \mathbb{C}^q,0$ is also free for any q. Given germs $V_i,0 \subset \mathbb{C}^{t_i},0$, i = 1,2, we can form their *product-union*

$$V_1 \uplus V_2 \overset{\text{def}}{=} V_1 \times \mathbb{C}^{t_2} \cup \mathbb{C}^{t_1} \times V_2 \subset \mathbb{C}^{t_1} \times \mathbb{C}^{t_2}.$$

Proposition 3.1: *If* $V_i,0 \subset \mathbb{C}^{t_i},0$ $i = 1,2$ *are free divisors then*

i) $V_1 \uplus V_2,0 \subset \mathbb{C}^{t_1} \times \mathbb{C}^{t_2},0$ *is a free divisor;*

ii) $T_{\log}(V_1 \uplus V_2)_{(x_0,y_0)} = T_{\log}(V_1)_{(x_0)} \oplus T_{\log}(V_2)_{(y_0)}$

iii) if H_i *are good defining equations for* V_i*, then* $H_1 \cdot H_2$ *is a good defining equation for* $V_1 \uplus V_2$.

Proof: For the first part, we quite generally claim that for germs $V_1,0 \subset \mathbb{C}^n,0$ and $V_2,0 \subset \mathbb{C}^m,0$, if $\{\zeta_1, \ldots, \zeta_p\}$ and $\{\eta_1, \ldots, \eta_q\}$ are sets of generators for Derlog(V_1) and Derlog(V_2) then they can be viewed as vector fields on $\mathbb{C}^{n+m}$ by composition with projection onto either factor and $\{\zeta_1, \ldots, \zeta_p, \eta_1, \ldots, \eta_q\}$ is a set of generators for Derlog($V_1 \uplus V_2$). Then, for V_1 and V_2 free divisors Derlog($V_1 \uplus V_2$) has n+m generators and is free by [Sa].

For the claim, we let $\mathbb{C}^n$ and $\mathbb{C}^m$ have local coordinates $\mathbf{x} = (x_1, \ldots, x_n)$ and $\mathbf{y} = (y_1, \ldots, y_m)$. Then,

$$\mathrm{Derlog}(V_1 \times \mathbb{C}^m) = \mathcal{O}_{\mathbb{C}^{n+m},0}\{\zeta_1, \dots, \zeta_p, \frac{\partial}{\partial y_1}, \dots, \frac{\partial}{\partial y_m}\} \quad \text{and}$$

$$\mathrm{Derlog}(\mathbb{C}^n \times V_2) = \mathcal{O}_{\mathbb{C}^{n+m},0}\{\frac{\partial}{\partial x_1}, \dots, \frac{\partial}{\partial x_n}, \eta_1, \dots, \eta_q\}.$$

Thus,

$$\begin{aligned} \mathrm{Derlog}(V_1 \circledast V_2) &= \mathrm{Derlog}(V_1 \times \mathbb{C}^m) \cap \mathrm{Derlog}(\mathbb{C}^n \times V_2) \\ &= \mathcal{O}_{\mathbb{C}^{n+m},0}\{\zeta_1, \dots, \zeta_p, \eta_1, \dots, \eta_q\}. \end{aligned} \tag{3.2}$$

Then, ii) follows from (3.2).

For the third claim, if H_i are good defining equations for V_i then there are germs of vector fields on $\mathbb{C}^n$ and $\mathbb{C}^m$ denoted e_i so that $e_i(H_i) = H_i$. Again both H_i and e_i can be viewed as defined on $\mathbb{C}^{n+m}$ by composition with projections onto the factors. Then, for $H = H_1 \cdot H_2$ and $e = 1/2(e_1 + e_2)$,

$$e(H) = 1/2(e_1 + e_2)(H_1 \cdot H_2) = 1/2(e_1(H_1) \cdot H_2 + H_1 \cdot e_2(H_2)) = H. \quad \square$$

Remark 3.3: If $\zeta_1 = e_1$ and $\eta_1 = e_2$ and the $\mathrm{Derlog}(H_i)$ are generated by $\{\zeta_2, \dots, \zeta_n\}$ respectively $\{\eta_2, \dots, \eta_m\}$ then the above proof shows that for $H = H_1 \cdot H_2$, $\mathrm{Derlog}(H)$ is generated by $\{\zeta_1 - \eta_1, \zeta_2, \dots, \zeta_n, \eta_2, \dots, \eta_m\}$.

Similarly, we can define the product-union $\circledast\ V_i$ for $i = 1 \dots q$ of q germs $V_i, 0 \subset \mathbb{C}^{t_i}, 0$, and an inductive application of the proposition implies the following corollary.

Corollary 3.4: *If each* $V_i, 0 \subset \mathbb{C}^{t_i}, 0$ $i = 1, \dots, q$ *are free divisors, then so is their product-union* $\circledast\ V_i$; *and if* H_i *is a good defining equation for* V_i, *then* $\prod H_i$ *is a good defining equation for* $\circledast\ V_i$.

Examples : By Mather's characterization of stability via multitranversality, the discriminant of a stable multi-germ is the product union of the discriminants of stable germs. Also, the "product" of arrangements is a product union. For example, the product-union of $\{0\} \subset \mathbb{C}$ with itself n times is the Boolean arrangement in $\mathbb{C}^n$ defined by $\prod z_i = 0$.

In general, neither of these results hold for almost free divisors; if $V,0 \subset \mathbb{C}^t,0$ is a almost free then $V \times \mathbb{C}^q,0$ will not in general be almost free, and similarly for product unions. However, there is a surprising consequence of proposition 3.1 for almost free divisors.

Definition 3.5: germs $V_i,0 \subset \mathbb{C}^n,0$ $i = 1,2$ will be said to be *algebraically transverse off 0* if for all x_0 in a punctured neighborhood of 0

$$T_{\log}(V_1)_{(x_0)} + T_{\log}(V_2)_{(x_0)} = T_{x_0}\mathbb{C}^n$$

Again we denote this by writing $V_1 \mathring{\pitchfork}_{\text{alg}} V_2$.

We can also define $V_i,0 \subset \mathbb{C}^n,0$ $i = 1,2$ to be *geometrically transverse off 0* if the strata of the canonical Whitney stratifications are transverse in a punctured neighborhood of 0; and we write $V_1 \mathring{\pitchfork}_{\text{geom}} V_2$.

Proposition 3.6: *If* $V_i,0 \subset \mathbb{C}^n,0$ $i = 1,2$ *are almost free divisors and* $V_1 \mathring{\pitchfork}_{\text{alg}} V_2$ *then* $V_1 \cup V_2,0 \subset \mathbb{C}^n,0$ *is an almost free divisor.*

Proof: Let $V_i,0 \subset \mathbb{C}^n,0$ be defined by $f_i: \mathbb{C}^n,0 \longrightarrow \mathbb{C}^{p_i},0$ as $V_i = f_i^{-1}(V_i')$ for free divisors $V_i',0 \subset \mathbb{C}^{p_i},0$. We define $f: \mathbb{C}^n,0 \longrightarrow \mathbb{C}^{p_1+p_2},0$ by $f(x) = (f_1(x), f_2(x))$. We easily see that $V_1 \cup V_2 = f^{-1}(V_1' \circledast V_2')$; and by proposition 3.1, $V_1' \circledast V_2'$ is a free divisor. Thus, it's enough to show $f \mathring{\pitchfork}_{\text{alg}} (V_1' \circledast V_2')$. Hence, we must show for all x_0 in a punctured neighborhood of 0 with $f(x_0) = y_0$,

$$df(T_{x_0}\mathbb{C}^n) + T_{\log}(V_1' \uplus V_2')_{y_0} = T_{y_0}\mathbb{C}^{p_1+p_2};$$

By ii) of proposition 3.1, this is equivalent to proving the surjectivity of

$$(3.7) \quad df(x_0)\colon T_{x_0}\mathbb{C}^n \longrightarrow T_{y_0}\mathbb{C}^{p_1}/T_{\log}(V_1')_{(y_0)} \oplus T_{y_0}\mathbb{C}^{p_2}/T_{\log}(V_2')_{(y_0)}$$

As each $f_i \mathring{\pitchfork}_{alg} V_i'$, the map to each factor is surjective, and so the surjectivity follows if $x_0 \notin V_1 \cap V_2$. Otherwise, by (2.12), $T_{\log}(V_i)_{(x_0)} = df_i^{-1}(T_{\log}(V_i')_{(y_0)})$. Since the algebraic transversality of the V_i implies $T_{x_0}\mathbb{C}^n = T_{\log}(V_1)_{(x_0)} + T_{\log}(V_2)_{(x_0)}$, the surjectivity of (3.7) follows by standard transversality arguments at the level of tangent spaces. □

This has a natural extension for general position of varieties.

Definition 3.8: The germs $V_i, 0 \subset \mathbb{C}^n, 0$, $i = 1, \dots, q$, will be said to be in *algebraic general position off 0* if the subspaces $T_{\log}(V_i)_{(x)}$, $i = 1, \dots, q$, are in general position in $T_x\mathbb{C}^n$ for all x in a punctured neighborhood of 0.

Then, there is the corollary of the proof of proposition 3.6.

Corollary 3.9 ("transverse union"): *If the germs* $V_i, 0 \subset \mathbb{C}^n, 0$, $i = 1, \dots, q$, *are almost free divisors and are in algebraic general position off 0, then*

$$\bigcup_{i=1}^{q} V_i, 0 \subset \mathbb{C}^n, 0 \text{ is an almost free divisor.}$$

Example 3.10: Let $V_i, 0 \subset \mathbb{C}^n, 0$, $i = 1, \dots, q$, be isolated hypersurface singularities defined by germs $h_i\colon \mathbb{C}^n, 0 \to \mathbb{C}, 0$. If for all $\mathbf{j} = \{j_1, \dots, j_r\}$, the germ $h_{\mathbf{j}} = (h_{j_1}, \dots, h_{j_r})\colon \mathbb{C}^n, 0 \to \mathbb{C}^r, 0$ defines an isolated complete intersection

singularity then $\bigcup_{i=1}^{q} V_i, 0 \subset \mathbb{C}^n, 0$ is an almost free divisor. In fact, $T_{\log}(V_i)_{(x)} = \ker(dh_i(x))$, and hence they are in general position exactly when for those j_1 with x

$\in V_{j_i}$, $i = 1, \ldots, r$, $\ker(dh_{j_i}(x))$ are in general position, i.e. $h_j = (h_{j_1}, \ldots, h_{j_r})$ defines an ICIS. We will also see in §6 that $\bigcup_{i=1}^{q} V_i, 0 \subset \mathbb{C}^n, 0$ can be viewed as an almost free nonlinear arrangement based on the Boolean arrangement.

Finally, we can see that almost free divisors behave well under pull-backs by finite map germs.

Proposition 3.11: *If* $V,0 \subset \mathbb{C}^t,0$ *is an almost free divisor and* $f_0\colon \mathbb{C}^s,0 \to \mathbb{C}^t,0$ *is a finite map germ with* $f_0 \mathrel{\mathring{\pitchfork}_{alg}} V$ *then* $f_0^{-1}(V), 0$ *is an almost free divisor.*

Proof: Let V be defined by $g_0^{-1}(V')$ where $V',0 \subset \mathbb{C}^q,0$ is free, $g_0\colon \mathbb{C}^t,0 \to \mathbb{C}^q,0$, and $g_0 \mathrel{\mathring{\pitchfork}_{alg}} V'$. By corollary 2.15, $g_0 \circ f_0 \mathrel{\mathring{\pitchfork}_{alg}} V'$. Hence, $f_0^{-1}(V) = g_0 \circ f_0^{-1}(V')$ is an almost free divisor. □

§4 Singular Milnor Fibers and Higher Multiplicities

Teissier [Te] defined a series of higher multiplicities for the case of isolated hypersurface singularities; these multiplicities are exactly the μ_i appearing in his μ^*-sequence $\mu^* = (\mu_0, \dots, \mu_n)$. Given $f_0 : \mathbb{C}^n,0 \longrightarrow \mathbb{C},0$, if P is a generic s-dimensional subspace in $\mathbb{C}^n$ then $f_0| P$ has an isolated singularity and he defines $\mu_s(f_0) = \mu(f_0| P)$, where μ denotes the usual Milnor number.

We wish to define analogous higher multiplicities for the case of nonisolated hypersurface singularities $V,0 \subset \mathbb{C}^n,0$. By the same stratification argument used by Teissier in the isolated case, a Zariski open subset of s-dimensional subspaces $P \subset \mathbb{C}^n$ are geometrically transverse to V off 0. However, if $f_0 : \mathbb{C}^n,0 \longrightarrow \mathbb{C},0$ is a defining equation for V, then $f_0| P$ will not in general have an isolated singularity. Lê and Teissier [LêT] considered instead the generic projections $V,0 \longrightarrow \Pi,0$ for linear subspaces Π of varying dimensions. They consider the Euler characteristics of the Milnor fiber for such projections, which they call the "vanishing Euler characteristics", and relate these numbers to the properties of the polar curves.

An alternate approach to this problem is to instead view the inclusion $i: P \hookrightarrow \mathbb{C}^n$ as a section of V. For simplicity we assume $P = \mathbb{C}^s$ so that we have a map germ $i: \mathbb{C}^s,0 \hookrightarrow \mathbb{C}^n,0$ which is geometrically transverse to V off 0 We begin by applying results of [DM].

Let $V,0 \subset \mathbb{C}^p,0$ be a free divisor with good defining equation H and $f_0 : \mathbb{C}^s,0 \longrightarrow \mathbb{C}^p,0$ a germ which is algebraically (resp. geometrically) transverse off 0. Let f_t be a (topological) stabilization i.e. for $t \neq 0$ f_t is algebraically transverse to V (resp. geometrically transverse) in an entire neighborhood of 0. Also, let B_ε denote a ball of radius $\varepsilon > 0$ about 0. Then, using a result of Lê [Lê1] [Lê2] the following was proven in the case $V' = f_0^{-1}(V)$ is finitely defined (i.e. $\nu_V(f_0) = \mathcal{K}_{H,e}\text{-codim}(f_0) < \infty$).

Theorem 4.1: ([DM,thm 5]) *For* $|t|$ *and* $\varepsilon > 0$ *sufficiently small,* $f_t^{-1}(V) \cap B_\varepsilon$ *is independent of the stabilization* f_t *and is homotopy equivalent to a bouquet of real* $(s-1)$*-spheres; and if* $s < \mathcal{h}(V)$, *the number of such spheres, denoted* $\mu_V(f_0)$ *equals* $\nu_V(f_0)$.

Furthermore, $f_t^{-1}(V) \cap B_\varepsilon$ is called the *singular Milnor fiber* of f_0; we shall also refer to it as the *singular Milnor fiber* of V'. Also, $\mu_V(f_0)$, which can be calculated by $\nu_V(f_0)$ in the above situation, is the *singular Milnor number*.

Remark: If $s \geq \mathcal{h}(V)$ in the above theorem, there is a correction factor which must be computed (see [DM, thm.6]). In the case that f_0 is only geometrically transverse to V in a punctured neighborhood of 0, the same theorem shows that for a topological stabilization of f_0, i.e. where f_t is geometrically transverse in an entire neighborhood of 0, the singular Milnor fiber of f_0 is still well-defined and is homotopy equivalent to a bouquet of spheres except now there is no known formula for the number of spheres $\mu_V(f_0)$.

Next, we see that as result of the good behavior of almost free divisors under pull-back by finite map germs, the preceding theorem extends to them. Let $V',0 \subset \mathbb{C}^n,0$ be an almost free divisor based on $V,0 \subset \mathbb{C}^p,0$ via $V' = g_0^{-1}(V)$ for $g_0 : \mathbb{C}^n,0 \longrightarrow \mathbb{C}^p,0$ with $\nu_V(g_0) < \infty$.

Corollary 4.2: *Suppose that* $f_0 : \mathbb{C}^s,0 \longrightarrow \mathbb{C}^n,0$ *is a finite map germ with* $s < n, \tilde{\mathcal{h}}(V)$ *which satisfies* $\nu_V(g_0 \circ f_0) < \infty$. *The singular Milnor number for* f_0 *(which equals the number of (s-1)- spheres in* $f_t^{-1}(V') \cap B_\varepsilon$ *for a stabilization* f_t*) is given by* $\mu_{V'}(f_0) = \nu_V(g_0 \circ f_0)$.

Proof: Let $f_t : \mathbb{C}^s,0 \longrightarrow \mathbb{C}^n,0$ be a stabilization of f_0. Then, as $s < n$, we may arrange $f_t(\mathbb{C}^s) \cap \{0\} = \emptyset$. Thus, by the proof of corollary 2.15, $g_0 \circ f_t$ is algebraically transverse to V' in a neighborhood of 0 iff f_t is algebraically transverse to V, which it is. Hence, $g_0 \circ f_t$ is a stabilization of $g_0 \circ f_0$. Thus, the singular Milnor fiber of $g_0 \circ f_0$ equals that for f_0. Hence, the result follows from the theorem 4.1 above. □

Now let i_t: $\mathbb{C}^s,0 \hookrightarrow \mathbb{C}^n,0$ be a (topological) stabilization of i; this means that i_t is algebraically (resp. geometrically) transverse to V in a neighborhood of 0. The singular Milnor fiber for i is $i_t(\mathbb{C}^s) \cap V \cap B_\varepsilon$, for B_ε a ball of radius $\varepsilon > 0$ and ε and t sufficiently small. We let $\mu_s(V)$ denote this singular Milnor number. We refer to $\{\mu_s(V)\}$ as the *set of higher multiplicities for* V (by definition, $\mu_0(V) = 1$). At times, for simplicity we denote $\mu_n(V)$, which is the singular Milnor number, by $\mu(V)$. These numbers can be computed in the case of almost free divisors.

Proposition 4.3: *Suppose that* V′ *is an almost free divisor based on* V *via* g_0. *Consider* i: $\mathbb{C}^s,0 \hookrightarrow \mathbb{C}^n,0$ *where* $s < \tilde{h}(V')$. *Suppose* i *actually satisfies the stronger condition that* $\nu_V(g_0 \circ i)$ *is finite and minimum among all nearby linear embeddings (in particular, as* $s < \tilde{h}(V')$, $i \mathrel{\mathring{\pitchfork}}_{geom} V' \Rightarrow \nu_V(g_0 \circ i)$ *is finite), then*

$$\mu_s(V) = \nu_V(g_0 \circ i).$$

Proof: Let $\mathcal{U}$ denote an open subset of Hom($\mathbb{C}^s$, $\mathbb{C}^n$) containing i. A deformation φ: $\mathbb{C}^s \times \mathcal{U}, (0, i) \longrightarrow \mathbb{C}^n,0$ of $g_0 \circ i$ is defined by $\varphi(x, i') = g_0 \circ i'(x)$. By [DM, prop. 5.2] $\mathcal{K}_{H,e}$-equivalence has what is called a "free deformation theory". This means that in a deformation such as φ, there is a neighborhood U of 0, and a neighborhood of i which we can assume is $\mathcal{U}$, such that for $i' \in \mathcal{U}$, the following expression is independent of i'

(4.4) $\sum \dim_{\mathbb{C}} N\mathcal{K}_{H,e}(g_0 \circ i', x)$ (summed over the finite number of non $\mathcal{K}_H$-stable points $x \in U$).

By assumption, $\nu_V(g_0 \circ i)$ (= $\dim_{\mathbb{C}} N\mathcal{K}_{H,e}(g_0 \circ i, 0)$) is minimum for nearby i'; thus, for all i' near i, there is only one $\mathcal{K}_H$-unstable point, namely 0, and $\nu_V(g_0 \circ i')$ is locally constant. Hence, some nearby i' is generic and $\nu_V(g_0 \circ i') = \nu_V(g_0 \circ i)$. The first part now follows immediately from corollary 4.2. The parenthetical statement follows from corollary 2.16. □

Remark: We have several comments to make regarding these multiplicities.

1) The condition that $\nu_V(g_0 \circ i)$ be minimum is really necessary as algebraic transversality off 0 does not imply genericity.

2) If $f_0 : \mathbb{C}^n,0 \longrightarrow \mathbb{C},0$ defines an isolated hypersurface singularity then $V' = f_0^{-1}(0)$ is an almost free divisor based on {0} and its "singular Milnor fiber" is exactly its Milnor fiber and $\nu_{\{0\}}(f_0) = \mu(f_0)$. By its definition, the higher multiplicity of an isolated hypersurface singularity case is exactly the Milnor number of the restriction. Thus, this sequence of higher multiplicites agrees with the μ^*-sequence in the isolated singularity case. Recall that the isolated singularity case satisfies $\tilde{h}(V') = \infty$ so the condition of proposition 4.3 is always satisfied.

3) Furthermore, in the case when $s = gm(V)-1$, the multiplicity $\mu_s(V)$ is exactly the vanishing Euler characteristic for the complex link of V in the sense of Goresky-MacPherson [GM].

4) Lastly, these multiplicities agree, up to a sign and the addition of 1, with the "vanishing Euler characteristics" of Lê-Teissier because the fibers of a generic projection form a stabilization of the generic section in our sense.

The preceding arguments indicate that we can compute some of the higher multiplicities even in the non-almost free case.

Definition 4.5: A hypersurface $(V',0) \subset \mathbb{C}^n,0$, is *free in codimension k* based on a free divisor $(V,0) \subset \mathbb{C}^p,0$, if there exists a germ $f_0: \mathbb{C}^n,0 \longrightarrow \mathbb{C}^p,0$ such that $V' = f_0^{-1}(V)$, and a complex analytic subset $W \subset V'$ of codimension k in $\mathbb{C}^n$ so that f_0 is algebraically transverse to V off W. We refer to f_0 as being *algebraically transverse to* V *in codimension k.* Likewise, we will refer to varieties being *algebraically transverse or in general position in codimension k* if the appropriate conditions hold off a complex analytic subset of codimension k.

Then we have the following corollary of the proof of 4.3.

Corollary 4.6: *Suppose that* V′ *is free in codimension k (based on* V*) via* g_0, *that* $i: \mathbb{C}^s,0 \hookrightarrow \mathbb{C}^n,0$ *satisfies the conditions of proposition 4.3, and in addition* $s < k$, *then* $\mu_s(V) = \nu_V(g_0 \circ i)$.

We mention one consequence of the proof of proposition 3.6 for varieties being free to codimension k.

Corollary 4.7: *Suppose that* $V'_i \subset \mathbb{C}^n,0$ *are free in codimension k (based on the free divisors* V_i *) for* $i = 1,\dots, r$. *If the* V'_i *are in algebraic general position in codimension k then* $\cup V'_i$ *is free in codimension k based on* $\uplus\, V_i$.

We consider several examples.

Examples (4.8):

a) The arrangement A' defined by $xyz(ax + by + cz) = 0$,$abc \neq 0$, is generic and is the pull back of the Boolean arrangement A via $\varphi: \mathbb{C}^3 \longrightarrow \mathbb{C}^4$ given by $\varphi(x, y, z) = (x, y, z, ax + by + cz)$. The singular Milnor fiber is the noncentral arrangement A'' defined by $xyz(ax + by + cz-t) = 0$ which equals $\varphi_t^{-1}(A)$ where φ_t is the stabilization $\varphi_t(x, y, z) = (x, y, z, ax + by + cz-t)$ which moves the image off the origin. We see there is one vanishing singular 2-cycle given by the (boundary of the) tetrahedron in fig. 4.9. That $\nu_A(\varphi) = 1$ follows from a direct calculation involving the vector fields in Derlog(A); however, we shall see in §5 a more general formula for it.

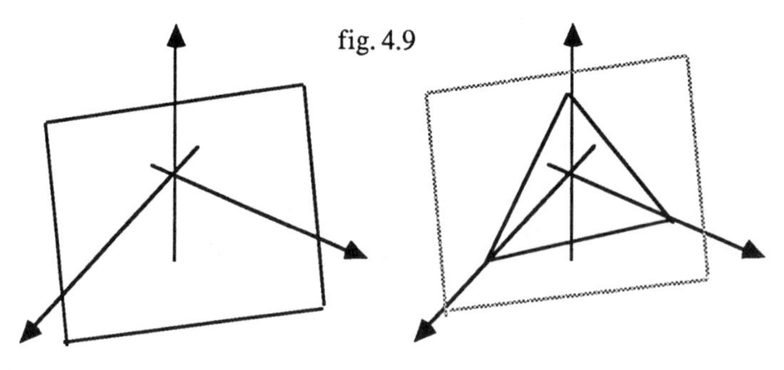
fig. 4.9

b) <u>Simple $\mathcal{A}_e$-codimension 1 germs</u>

In [DM] was considered the unfoldings, versal except for the Hessian deformation, of the simple elliptic or exceptional unimodal hypersurface singularities. By the results of [L] and [W], although they have $\mathcal{A}_e$-codimension 1, they only deform in a topologically trivial fashion so the $\mathcal{A}_e$-codimension is not exhibited in the singular Milnor fiber of the discriminant. However, the same calculations of Looijenga but applied to the simple hypersurface singularities show that unfoldings, versal except for the Hessian deformation, still have $\mathcal{A}_e$-codimension = 1. Now when we deform by the Hessian term we stabilize the germ, and this is not topologically trivial. When we examine the discriminant under this stabilization, we see a single singular (μ-2)-cycle "bubbling off" from the origin, as in fig. 4.10 . In fact, this is the correct picture for the discriminants of all simple hypersurface singularities.

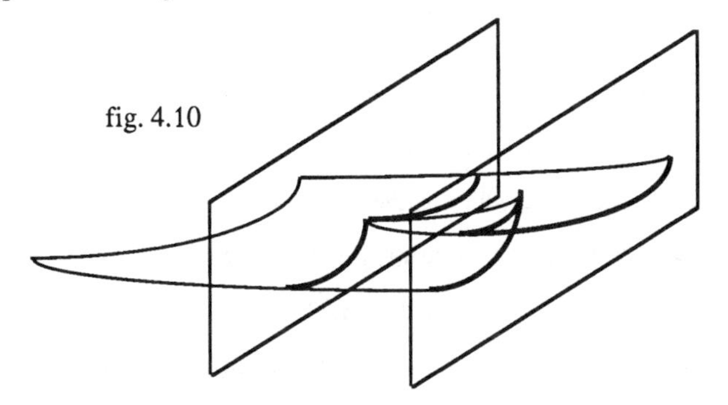
fig. 4.10

These are not all of the $\mathcal{A}_e$-codimension 1 germs even among those obtained by unfolding simple hypersurface singularities. If $\{\varphi_0=1, \varphi_1, \ldots, \varphi_q\}$ is a weighted homogeneous basis for the Jacobian algebra of the simple germ $f_0(x)$ with φ_q denoting the Hessian, then the germ $F : \mathbb{C}^{n+q-1+p},0 \longrightarrow \mathbb{C}^{q+p},0$

$$F(x, u, v) = (f_0(x) + \sum_{i=1}^{q-1} u_1\varphi_i + Q(v)\varphi_q , u, v) ,$$

where Q(v) is a nonsingular quadratic form, again has $\mathcal{A}_e$-codimension 1 and its discriminant has a stabilization creating a single q+p−1 cycle as above. In general, the question of $\mathcal{A}_e$-codimension 1 germs which are not unfoldings of hypersurface germs becomes more subtle; David Mond has informed this author that these have been classified by Victor Goryunov [Go]. In [Mo2] the corresponding picture for the REALization of the complex singular cycles for the real singularities is discussed; see also §§ 10 and 11 for examples of nonREALizable singular complex cycles.

c) Free to Codimension k

For a germ $f : \mathbb{C}^s,0 \longrightarrow \mathbb{C}^t,0$ defining an ICIS, we let $F: \mathbb{C}^{s+q},0 \longrightarrow \mathbb{C}^{t+q},0$ denote a stable unfolding. The versality discriminant of f, V(f), is the set of points in $\mathbb{C}^t$ where f, viewed as a deformation, is not versal (see [D3]). If k = codim(V(f)), then D(f) is free in codimension k when viewed as a section of D(F) via $i: \mathbb{C}^t,0 \hookrightarrow \mathbb{C}^{t+q},0$ (this follows from the proof of prop. 2.3 of [D2]). Hence, if $j < \text{codim}(V(f))$, then $\mu_j(D(f)) = \min \nu_{D(F)}(i')$ where $i' = i \circ j$ and $j: \Pi \hookrightarrow \mathbb{C}^t$ is a linear embedding of a dimension j subspace Π. If $s < t$, then D(F) = Image(F) is not free in general. For example, $F(x, y) = (x^2, xy, y)$ has image the Whitney umbrella $X^2Y = Z^2$. It is free in codimension 2, as are the "twisted Whitney umbrellas" $(x^2, xg(y, x^2), y)$ obtained as pullbacks. However, this codimension restriction leaves untouched the important case of $\mu_3(D(f))$, which is the singular Milnor number of D(f) and has the surprising relation with $\mathcal{A}_e$-codim(f) [Mo1] and [JS].

The arrangement A of hyperplanes in $\mathbb{C}^4$ given by

$$Q(x, y, z, w) = xyzw(x+y-z)(x-2y+z)(2x+z)(x+y+z+w) = 0$$

is a union of two Boolean arrangements (which are free)

$$xyzw = 0 \quad \text{and} \quad (x+y-z)(x-2y+z)(2x+z)(x+y+z+w) = 0.$$

These arrangements fail to be algebraically ($\Longleftrightarrow$ geometrically) transverse on the w-axis. Hence, by corollary 4.7, their union is free in codimension 3. Hence, we can still compute $\mu_j(A)$ for $j < 3$ as $\nu_A(\varphi \circ i)$, where $i: \Pi \hookrightarrow \mathbb{C}^4$ is a generic j-dimensional subspace and φ defines A as an arrangement based on a product of Boolean arrangements. We shall see in § 5 that, e.g., $\mu_2(A) = \binom{7}{2} = 21$.

Part II Linear and Nonlinear Arrangements

We will consider in part II arrangements of hyperplanes as well as nonlinear arrangements of hypersurface singularities. We shall see that basic invariants which play a key role in the topology of complements of arrangements are the higher multiplicities. For an A-generic arrangement $A' \subset \mathbb{C}^n$ (with A free), we compute the higher multiplicities of A' in terms of the Macaulay-Bezout numbers defined from $\exp'(A)$, which is a slightly modified form of the exponents of A (prop. 5.2). Second, we prove quite generally (prop. 5.6) that the Folkman complex of any arrangement A' can be embedded as a strong deformation retract of the singular Milnor fiber for a generic hyperplane section. From this it follows that $\mu_{n-1}(A')$ equals $|\mu(A')|$, the absolute value of the Mobius function applied to A'. This allows us to give a formula for the Betti numbers for the complement of A' in terms of the higher multiplicities (lemma 5.8). Combining this formula with the computations for the Macaulay-Bezout numbers given in [D4] yields an algebraic formula for the Poincaré polynomial for A-generic arrangements (theorem 1). In the case of free arrangements this recovers (and gives a new proof of) the factorization theorem of Terao.

For nonlinear arrangements, we likewise give in § 6 algebraic formulas for the higher multiplicities. Again in the homogeneous or weighted homogeneous case, these formulas can be computed using Macaulay-Bezout numbers.

§5 Almost Free Arrangements

One of the basic questions concerning arrangements of hyperplanes $A \subset \mathbb{C}^n$ is to determine the topology of the complement $M(A) = \mathbb{C}^n \backslash A$. There is a special class of free arrangements for which Terao has given an especially elegant formula for the Poincaré polynomial $P(A, t)$ of $M(A)$. Because A, thought of as a singular hypersurface singularity, is free, Derlog(A) is a free $\mathcal{O}_{\mathbb{C}^n,0}$-module of rank n. Furthermore, because the defining equation Q for A (a product of linear factors defining the hyperplanes in A) is homogeneous, the generators of Derlog(A) may be chosen to be homogeneous. Their degrees as vector fields form the degrees of A, $\deg(A) = (d_1, \dots, d_n)$, with the d_i given in ascending order. Since the Euler vector field is always in Derlog(A) one of the $d_i = 0$ and Q is a good defining equation for A (also if A is not essential then there are $d_i = -1$ corresponding to the factor T). There are also the exponents $\exp(A) = (e_1, \dots, e_n)$ where $e_i = d_i+1$ denotes the degree of the coefficient functions for the i-th generator. Then, Terao's factorization theorem,which encompasses earlier work of Arnold, Brieskorn, and Orlik-Solomon, gives an algebraic formula for $P(A, t)$

$$P(A, t) = \prod_{i=1}^{n} (1 + e_i t)$$

In this section we will give a new proof of this which will also extend the result to A-generic arrangements using the higher multiplicities for A. From §1 we recall that A′ is an almost free arrangement based on $A \subset \mathbb{C}^p$ if $A' = \varphi^{-1}(A)$ for a linear 1-1 map $\varphi: \mathbb{C}^n \longrightarrow \mathbb{C}^p$ which is algebraically transverse off 0 to A.

Definition 5.1: An arrangement A' will be called *A-generic* if it is of the form $A' = B \times T$ for an almost free arrangement B based on A. We will say A' is *A-generic to codimension* ℓ if the linear map φ defining B is algebraically transverse to A to codimension ℓ.

Since $h(A) = \infty$, algebraic transversality is equivalent to geometric transversality; hence, $\varphi(\mathbb{C}^n)$ is transverse to all intersections of hyperplanes of A off 0. We shall refer to a k-dimensional subspace $K \subset \mathbb{C}^p$ as a *generic section of* A if K intersects transversally any $\cap H_i \neq (0)$ for $H_i \subset A$. We also denote a *codimension-k generic section* of A by $A^{(k)}$. Then, by the above, any almost free arrangement based on A is of the form $K \cap A$ for a generic section K. If K is a generic section, then $i: K \hookrightarrow \mathbb{C}^n$ is geometrically transverse to A' off 0; and since $\tilde{h}(A') = \infty$, by corollary 2.16 $K \cap A$ is finitely defined and so its singular Milnor fiber can be computed as $\nu_A(i \circ \varphi)$. Thus, for such arrangements we can compute the higher multiplicities as follows.

Proposition 5.2: *If A' an A-generic arrangement and* $\exp'(A)$ *denotes the p-1 tuple obtained by excluding a single* 1 *from* $\exp(A)$, *then for* $k < r = r(A')$

$$\mu_k(A') = B_k(\exp'(A)) = \sigma_k(\exp'(A)).$$

If A' an A-generic to codimension ℓ, then the above conclusions hold for $k < \ell$.

For an n-tuple $\mathbf{d} = (d_1, \ldots, d_n)$, $B_k(\mathbf{d})$ is the Macaulay-Bezout number defined in [D4], which is computed as $\sigma_k(\mathbf{d})$, the k-th elementary symmetric function in the d_i.

Remark: If $k \geq r$, then a generic section is actually transverse to A' and hence is just A' up to a trivial factor, as would be the Milnor fiber of a stabilization. This is similar to the construction of the complex link in [GM] where first a normal slice is taken before taking a codimension 1 section in that slice.

Proof: We can view $A' = B \times T$ where $B = K_1 \cap A$ for a generic section $K_1 \hookrightarrow \mathbb{C}^n$ of A. Then, if $i: K \hookrightarrow K_1$ is a generic k-dimensional section of B, then the composition with $j: K_1 = K_1 \times \{0\} \hookrightarrow K_1 \times T$ yields a generic k-dimensional section of A', $i': K \hookrightarrow \mathbb{C}^n$. Furthermore, a stabilization i_t of i composes with j to give a stabilization i'_t of i'. Thus, the singular Milnor Fibers of i and i' are the same. Hence, $\mu_k(A') = \mu_k(B)$. If we can apply proposition 4.3, then $\mu_k(B) = \nu_A(j \circ i)$. For this we must verify that $\nu_A(j \circ i)$ is minimal among $\nu_A(j \circ i'')$ for i'' a nearby linear embedding i″: $K \hookrightarrow K_1$. This follows from the calculation of $\nu_A(j \circ i)$.

Lastly, as $i' = j \circ i$ is a generic section of A, we can compute $\nu_A(i')$ using 2.7. We locally represent $i' : \mathbb{C}^k \longrightarrow \mathbb{C}^p$ with coordinates $(x_1, \ldots, x_k)$ for $\mathbb{C}^k$ and $(x_1, \ldots, x_p)$ for $\mathbb{C}^p$. Derlog(Q) has homogeneous generators $\{\zeta_1, \ldots, \zeta_{p-1}\}$ of degrees = deg(A) with a single 0 removed corresponding to the Euler vector field. Then

$$\begin{aligned}\nu_A(i') &= \dim_{\mathbb{C}} \mathcal{O}_{\mathbb{C}^k,0}\left\{\frac{\partial}{\partial x_1}, \ldots, \frac{\partial}{\partial x_p}\right\} / \left(\mathcal{O}_{\mathbb{C}^k,0}\left\{\frac{\partial}{\partial x_1}, \ldots, \frac{\partial}{\partial x_k}, \zeta_1 \circ i', \ldots, \zeta_{p-1} \circ i'\right\}\right) \\ &= \dim_{\mathbb{C}} \left(\mathcal{O}_{\mathbb{C}^k,0}\right)^{p-k} / \left(\mathcal{O}_{\mathbb{C}^k,0}\{F_1, \ldots, F_{p-1}\}\right).\end{aligned}$$

Here we represent the (p−k)-tuples of coefficient functions of $\left\{\frac{\partial}{\partial x_{k+1}}, \ldots, \frac{\partial}{\partial x_p}\right\}$ for the generators $\{\zeta_1 \circ i', \ldots, \zeta_{p-1} \circ i'\}$ as $\{F_1, \ldots, F_{p-1}\}$. Then, the coefficient functions of the generators of Derlog(Q) have degrees exp′(A). As $\nu_A(i')$ is finite (as $k < \tilde{h}(A') = \infty$) it equals $B_k(\exp'(A))$ by theorem 1 of [D4]. □

Example 5.3: We saw in §3 that the arrangement A defined by $xyz(ax + by + cz) = 0$,$abc \neq 0$, is generic (in fact, it is almost free) based on the Boolean arrangement $B \subset \mathbb{C}^4$. Thus, since exp(B) = (1, 1, 1 ,1), exp′(B) = (1^3) and $\mu_j(A) = B_j(1^3) = \binom{3}{j}$ so $\mu_3(A) = 1$, $\mu_2(A) = 3$, and $\mu_1(A) = 3$.

We have seen in example 4.8 the singular 2-cycle in the singular Milnor fiber for A Below we see the 3 singular 1-cycles and the 3 singular 0-cycles.

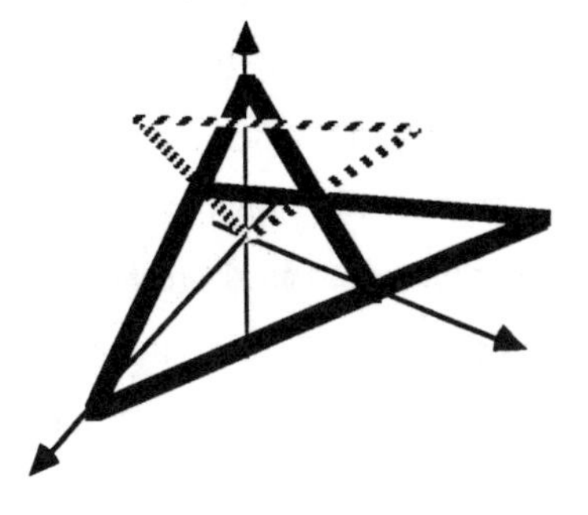

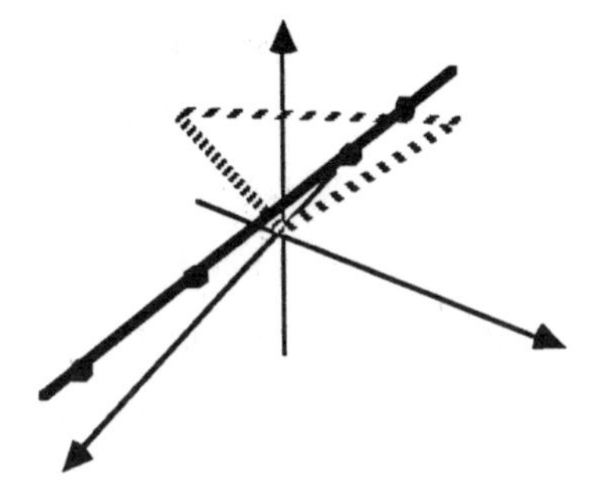

More generally if $A \subset \mathbb{C}^n$ is any generic arrangement in the sense of [OT,§5.1] consisting of p-hyperplanes in general position off 0 defined by the linear forms $\ell_i(x)$ then it is almost free based on a Boolean arrangement $B \subset \mathbb{C}^p$ and defined by the linear map $\varphi(x) = (\ell_1(x), \dots, \ell_p(x))$. Thus, for any $j < r(A)$,

$$\mu_j(A) = B_j(1^{p-1}) = \binom{p-1}{j}$$

Example 5.4: An arrangement A can be generic for two different free arrangements. The significance of this is that the top multiplicities differ reflecting the way that the singular Milnor fiber for A preserves the lattice structures of the free arrangements. For example, if A = {6 lines through 0 in $\mathbb{C}^2$ }, then (in addition to A itself being free) it is also generic based on the Boolean arrangement in $\mathbb{C}^6$. Its singular Milnor fiber is the noncentral generic arrangement of 6 lines in $\mathbb{C}^2$. We can see the $\mu_2(A) = B_2(1^5) = \sigma_2(1^5) = 10$ singular 1-cycles in fig. 5.5a.

Also, A is also generic for the Braid arrangement $B' \subset \mathbb{C}^4$ (defined by

$\prod_{i \neq j}(y_i - y_j))$ via a linear inclusion $\varphi': \mathbb{C}^2 \longrightarrow \mathbb{C}^4$. Then, $\exp(B') = (0, 1, 2, 3)$ (note $r(B') = 1$). Via φ' the singular Milnor fiber will preserve the lattice structure and will have only the $B_2(2, 3) = \sigma_2(2, 3) = 6$ singular 1-cycles in fig 5.5b.

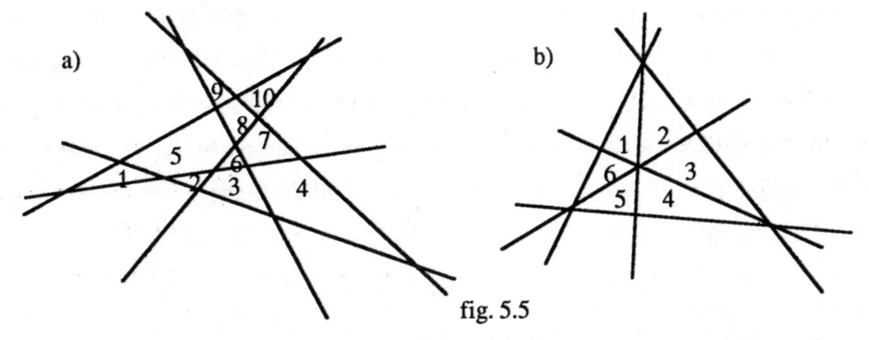

fig. 5.5

Note: Despite the difference for the top multiplicities, the lower ones will be the same and are independent of the free divisor.

Next we see that these higher multiplicities are related to the Poincare polynomial $P(A', t)$ of $M(A')$ for an A-generic arrangement $A' \subset \mathbb{C}^n$. There is associated to the arrangement A' the Folkman complex $F(A')$ [F] [OT,4.5], which is a simplicial complex which has the homotopy type of a bouquet of spheres of real dimension $r(A')-2$. The number of spheres equals $|\mu(A')|$ where μ denotes the Mobius function for the arrangement.

Proposition 5.6: *Suppose that $A' \subset \mathbb{C}^n$ is an essential arrangement and $i: H' \hookrightarrow \mathbb{C}^n$ is a generic hyperplane section to A'. There is an embedding of $F(A')$ as a deformation retract into the singular Milnor fiber of i. Hence,*

$$|\mu(A')| = \nu_A(i) = \mu_{n-1}(A').$$

Remark: Falk [Fa] [OT,§5.6] constructed an embedding of $F(A')$ into $A' \cap S^{2n-1}$ which is a homotopy equivalence up through dimension $n-2$. This proposition shows that the Folkman complex provides an exact homotopy model for the singular Milnor fiber of a generic hyperplane section.

Proof: That $F(A')$ is homotopy equivalent to the singular Milnor fiber is essentially given by Goresky-MacPherson [GM, Part III,chap. 1]. Given any arrangement A' of affine subspaces of arbitrary dimension, they construct homotopy equivalences $\varphi : C(A') \longrightarrow \mathbb{R}^n$ and $\pi : C(A') \longrightarrow K(A')$, where the image of φ is A' and $K(A')$ is the order complex for the lattice associated to A'. Here $C(A') = \cup\, |\, v| \times K(A'_{\geq v})$ where the union is over the "flats" $|\, v| = \cap H_i$, for H_i hyperplanes of A', and $K(A'_{\geq v})$ is the order complex for the sublattice of $w \supseteq v$. Also, φ and π denote projections onto each factor.

The order complex can be explicitly embedded in the singular Milnor fiber of i as follows. First, H' is transverse to all one-dimensional flats $L_j = \cap H_i$, for H_i hyperplanes of A'. Pick a line distinct from these; a stabilization is given by translating H' t units along this line. We denote this translate by H'_t and the affine embedding by i_t. Then, $H'_t \cap L_i = \{p_i\}$ a point. Here we fix a small t. Suppose $p_{i_0}, \dots, p_{i_s}$ are in general position in H'_t. If there is an s-dimensional flat X containing these p_{i_j} then let σ_X denote the convex hull of these points and pick a point $p_X \in \mathrm{int}(\sigma_X)$.

We map $K(A')$ onto the barycentric subdivision of $\cup\sigma_X$. Given a sequence of flats $\tilde{X} : X_0 \subset X_2 \dots \subset X_k$ we map the corresponding k- simplex $\sigma_{\tilde{X}}$ of $K(A'\backslash\{0, \mathbb{C}^n\})$ to $\tilde{\sigma}_{\tilde{X}}$ the k-simplex spanned by $\{p_{X_0}, \dots, p_{X_k}\}$. This defines a continuous map

$$K(H'_t \cap A'\backslash\{H'_t\}) \longrightarrow C(H'_t \cap A'\backslash\{H'_t\})$$

$$(\sigma_{\tilde{X}}, q) \longmapsto ((\sigma_{\tilde{X}}, q), (\tilde{\sigma}_{\tilde{X}}, q)).$$

where q denote barycentric coordinates. This map is a left inverse to π. Since π is a homotopy equivalence, so is this map and hence, so is its composition with φ. It embeds $K(H'_t \cap A'\backslash\{H'_t\})$ as the barycentric subdivision. Since

$$K(A'\backslash\{0, \mathbb{C}^n\}) \simeq K(H'_t \cap A'\backslash\{H'_t\}),$$

and $K(A'\backslash\{0, \mathbb{C}^n\})$ is the Folkman complex for A' we obtain the desired embedding.

Lastly, on the one hand, $K(A'\backslash\{0, \mathbb{C}^n\})$ has the homotopy type of a bouquet of $|\mu(A')|$ spheres of real dimension $r(A')-2 = n-2$; while by theorem 3.1 it consists of $\nu_A(i) = \mu_{n-1}(A')$ spheres. Thus, these numbers agree. □

Remark: This says that $|\mu(A')|$ is the singular Milnor number for the generic hyperplane section of A' (a surprising coincidence of notation!). Notice that this formula works both ways. It gives an algebraic formula for $|\mu(A')|$, namely $\mu_{n-1}(A') = B_{n-1}(\exp'(A))$ in the case A' is A-generic for free A. However, it also gives a combinatorial formula for the singular Milnor number in the non-almost free (or non-generic) cases where we know no algebraic formula, and thus, provides a starting point for understanding the non-almost free hypersurfaces.

A consequence of the proof of the preceding proposition is a formula for the number of compact regions for a real noncentral arrangement, i.e. the number of connected components of the complement whose closures are compact. By a noncentral arrangement A we mean one for which the intersection of all of its hyperplanes is empty. In analogy with the central case we define

$$\mu(A) = -\sum \mu(X) \quad \text{summed over all affine flats X in } A$$

Then, we obtain a formula which is originally due to Zaslavsky [Z].

Corollary 5.7: *Let $A \subset \mathbb{R}^n$ be a non-central arrangement then*

$$\textit{the number of compact regions in } \mathbb{R}^n \setminus A = |\mu(A)|.$$

Proof: First, we have the associated complex arrangement $A_{\mathbb{C}}$. There is a standard procedure for obtaining from $A_{\mathbb{C}}$ a central arrangement A' in $\mathbb{C}^{n+1}$. For each hyperplane H in $A_{\mathbb{C}}$, we form the hyperplane $H' \subset \mathbb{C}^{n+1}$ through 0 which is the linear subspace spanned by H. The union of H′ gives A'. Also, we see that $\mathbb{C}^n \times \{0\}$ is a generic section of A' which has a singular Milnor fiber $A' \cap \mathbb{C}^n \times \{t\} \simeq A_{\mathbb{C}}$. Also, the preceding proof gave an explicit embedding of the Folkman complex of A' as a deformation retract of $A_{\mathbb{C}}$. This complex is actually embedded in A. Furthermore, the homotopy equivalence can be made to commute with conjugation so we obtain a strong deformation retraction $F(A') \simeq A_{\mathbb{C}}$. Hence,

$$\operatorname{rank}(H^{n-1}(A_{\mathbb{C}}; \mathbb{Z})) = \operatorname{rank}(H^{n-1}(F(A'); \mathbb{Z})) = |\mu(A')|.$$

Also, $F(A')$ is a compact subset of $\mathbb{R}^n$ and by Alexander duality

$$\operatorname{rank}(\tilde{H}_0(\mathbb{R}^n \setminus F(A'); \mathbb{Z})) = \operatorname{rank}(H^{n-1}(F(A'); \mathbb{Z})) = |\mu(A')| = |\mu(A)|.$$

Thus, when we exclude the one noncompact region of $\mathbb{R}^n \setminus F(A')$, we have $|\mu(A)|$ compact regions which is the same number for $\mathbb{R}^n \setminus A$. □

Via proposition 5.6 we are also able to give the Poincaré polynomial for an A-generic arrangement. It depends on the following observation.

Lemma 5.8: *Let $A \subset \mathbb{C}^n$ be an essential arrangement and let $a_i(A)$ denote the coefficient of t^i in the Poincaré polynomial $P(A, t)$. Then, for $0 \le i \le n-1$*

i) $$a_i(A) = \mu_i(A) + \mu_{i-1}(A).$$

ii) *in particular,*

$$a_{n-1}(A) = |\mu(A)| + |\mu(A^{(1)})|.$$

iii) *Thus,*

$$P(A, t) = (1+t)\cdot\mu(A, t)$$

where $\mu(A, t) = \sum_{j=0}^{n-1} \mu_j(A)t^j$ *is the "multiplicity polynomial" of A.*

Proof: First, we prove the special case ii) of the lemma. To establish it, we use the expression for $P(A, t)$ in terms of the Mobius function for the lattice $L(A)$ [OT,§5.4]. The coefficient of t^i is the sum of $(-1)^i\mu(X)$ over all flats X of codimension i. Hence,

$$a_{n-1}(A) = \sum (-1)^{n-1}\mu(L)$$

summed over 1-dimensional flats of A. On the other hand,

(5.9) $$\mu(A) = -\sum \mu(X) \quad \text{summed over all } X \neq (0).$$

If H is a hyperplane which defines a generic section of A, then every flat X of A of dimension > 1 will contribute a flat $X \cap H$ of $H \cap A$; moreover, by the inductive definition of μ, $\mu(X \cap H) = \mu(X)$ provided $\dim(X) > 1$. Thus, applying (5.9) to both A and $A' = H \cap A$,

(5.10) $$\mu(A) - \mu(A') = -\sum \mu(L) = (-1)^n a_{n-1}(A)$$

summed over 1-dimensional flats L of A. Since quite generally $\text{sign}(\mu(A)) = (-1)^{r(A)}$ [OT,§2.2], $\mu(A)$ and $\mu(A')$ have opposite signs. Thus, taking absolute values gives the result.

For the general case, let K be a k-dimensional generic section of A, and let $A' = K \cap A$. Then, $|\mu(A')| = \mu_{k-1}(A)$. Also, if H' is a generic section of A' in K, then by corollary 2.15, H' is a (k-1)-dimensional generic section of A. Hence, $\mu_{k-1}(A') = \mu_{k-1}(A)$. Finally, by repeating the above reasoning used for ii), $a_k(A) = a_k(A')$. The result follows by applying the special case ii) to A'. □

From this lemma together with proposition 5.2, we immediately obtain an extension of Terao's factorization theorem.

Theorem 1: *Let* $A' \subset \mathbb{C}^n$ *be an A-generic arrangement with A free. Then,* *with* $r = r(A')$ *(and* σ_j *denoting the j-th elementary symmetric function)*

$$P(A', t) = \sigma_{r-1}(\exp'(A))\cdot t^r + \sum_{j=0}^{r-1} \sigma_j(\exp(A))\cdot t^j$$

In the case of the free arrangement A itself, $\sigma_{j-1}(\exp'(A)) = \sigma_j(\exp(A))$ for $j \geq r(A)$ *and so*

$$P(A, t) = \prod_{i=1}^{n} (1 + e_i t) \qquad \textit{where } \exp(A) = (e_1, \dots, e_n).$$

Remark: The relation between $P(A, t)$ and $P(A', t)$ implied by theorem 1 also partially follows from the local Lefschetz theorem for singular varieties due to Hamm-Lê [HLê], which also applies to nonlinear arrangements.

Proof: It is sufficient to verify it for the case of A' a generic section of dimension n of an essential free arrangement A. Then, A' itself is essential so $n = r$ $(= r(A'))$ and by Lemma 5.8,

$$a_k(A') = \mu_k(A') + \mu_{k-1}(A') \qquad \text{if } k < n$$

and

$$a_n(A') = |\mu(A')| = \mu_{n-1}(A)$$

Also, by proposition 5.2

$$\mu_k(A') = B_k(\exp'(A)) = \sigma_k(\exp'(A)).$$

The result follows on observing

$$\sigma_k(\exp'(A)) + \sigma_{k-1}(\exp'(A)) = \sigma_k(\exp(A)) \qquad \square$$

As a corollary of the proof we obtain

Corollary 5.11: *Let* $A' \subset \mathbb{C}^n$ *be A-generic to codimension* ℓ, *then for* $j < \ell$

$$\mu_j(A) = \sigma_j(\exp'(A))$$

and

$$a_j(A) = \sigma_j(\exp(A))$$

Example 5.12: We considered in example 5.3 the generic arrangement A defined by $xyz(ax + by + cz) = 0$,$abc \neq 0$ based on the Boolean arrangement $B \subset \mathbb{C}^4$. Thus, since $\exp(B) = (1, 1, 1, 1)$, $\exp'(B) = (1^3)$ so $\mu_j(A) = B_j(1^3) = \binom{3}{j}$. Thus,

$$P(A, t) = \binom{3}{2} \cdot t^3 + \binom{4}{2} t^2 + \binom{4}{1} t + 1 = 3t^3 + 6t^2 + 4t + 1$$

More generally, if as in example 5.3, A is a generic arrangement of p planes in $\mathbb{C}^n$ with $r(A) = n$. Then, A is based on the Boolean arrangement $B \subset \mathbb{C}^p$, so $\exp(B) = (1^p)$ and

$$P(A, t) = \binom{p-1}{n-1} \cdot t^n + \sum_{j=0}^{n-1} \binom{p}{j} t^j$$

Example 5.13: We considered in example 4.8 c) the arrangement $A \subset \mathbb{C}^4$ defined by $xyzw(x+y-z)(x-2y+z)(2x+z)(x+y+z+w) = 0$. We observed that it is a union of two Boolean arrangements which are algebraically transverse in codimension 3. Corollary 4.7 implies A is free to codimension 3 based on

$B \uplus B$, where B again denotes the Boolean arrangement in $\mathbb{C}^4$. Hence, since

$\exp(B \uplus B) = (1^8)$, corollary 5.11 implies that

$$\mu_j(A) = B_j(1^7) = \binom{7}{j} \quad \text{and} \quad a_j(A) = B_j(1^8) = \binom{8}{j} \quad \text{for } j < 3.$$

Example 5.14: For the free arrangement $B_3 \subset \mathbb{C}^3$ defined by

$$Q = xyz(x-y)(x-z)(z-y)(x+y)(x+z)(z+y)$$

$\exp(B_3) = (1, 3, 5)$. If we take the 2-dimensional generic section A, then

$$P(B_3, t) = 15t^3 + 23t^2 + 9t + 1 = (1+t)(1+3t)(1+5t)$$

while

$$P(A, t) = \sigma_1(3, 5)t^2 + \sigma_1(1, 3, 5)t + 1 = 8t^2 + 9t + 1.$$

We remark that again we can see the multiplicity $\mu_1(A) = |\mu(A^{(1)})| = 8$ as the 8 singular 0-cycles obtained by intersecting the nine lines forming A by a generic line off the origin.

§6 Nonlinear Arrangements

In this section we consider nonlinear arrangements. There are several possible meanings for this term including: i) a union of nonlinear hypersurfaces, ii) the restriction of an arrangement of hyperplanes to a nonlinear subspace (singular or not), and iii) a combination of i) and ii) which is a union of hypersurfaces on a nonlinear subspace. What we shall do will ultimately include all of these possibilities; however, initially in this section, we shall concentrate on i). In §8, we will include the other two possibilities under the the heading of almost free complete intersections.

(6.1) Simple Examples (of Milnor Fibers) of Nonlinear Arrangements :

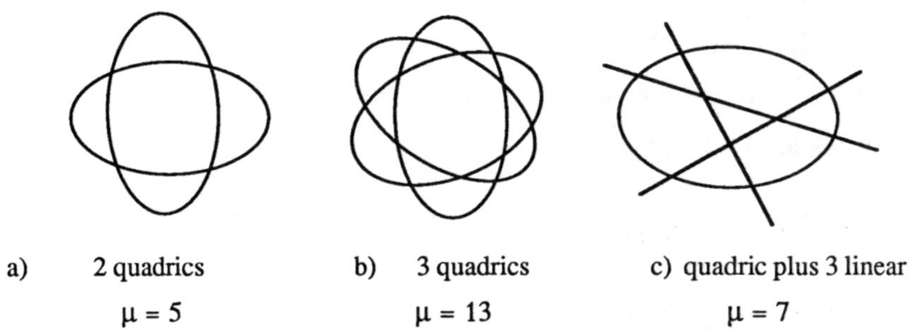

a) 2 quadrics $\mu = 5$ b) 3 quadrics $\mu = 13$ c) quadric plus 3 linear $\mu = 7$

For these nonlinear arrangements, we can see the singular cycles bounding the compact regions in the complement, just as for linear arrangements.

We shall apply results of §4 to give formulas for the singular Milnor numbers and higher multiplicities of nonlinear arrangements in terms of lengths of certain determinantal Cohen-Macaulay modules. We are especially interested in giving formulas in the weighted homogeneous case analogous to those for linear arrangements. This will use theorem 2 from [D4], which computes weighted Macaulay-Bezout numbers to give such a formula (Prop. 6.9) in terms of the τ-function. We will apply the formulas to: i) a "homogeneous nonlinear arrangement" of homogeneous hypersurfaces of fixed degree; ii) a "generic arrangement" of weighted homogeneous hypersurfaces of varying degrees; and iii) an arrangement which is a "mixture" i.e. a union of a linear arrangement with hypersurface singularities in general position.

It is much more difficult to compute the topology of the complement for nonlinear arrangements. Much of the topology can be deduced inductively by the singular local Lefschetz theorem of Lê-Hamm [LêH]; however, unlike the case of linear arrangements, the complete determination requires more than just the higher multiplicities and will be considered in a later paper.

We will use the results for nonlinear arrangements in §8 to compute the higher multiplicities for nonisolated complete intersections (which are almost free). We will do this by taking an unexpected dual approach to the topology of intersections of algebraically transverse varieties, namely, we derive information about intersections from information about their unions. These unions are special cases of nonlinear arrangements.

Singular Milnor Numbers of Nonlinear Arrangements

Let $A \subset \mathbb{C}^p$ be a free arrangement. We may choose a set of homogeneous generators $\{\zeta_0, \zeta_1, \dots, \zeta_{p-1}\}$ for Derlog(A), with ζ_0 denoting the Euler vector field, and so that for h a homogeneous defining equation for A, Derlog(h) is generated by $\{\zeta_1, \zeta_2, \dots, \zeta_{p-1}\}$. We let $\exp(A) = (e_0, \dots, e_{p-1})$ with $e_0 = 1$ and $e_1 \leq \dots \leq e_{p-1}$.

Definition 6.2: Suppose $f_0 : \mathbb{C}^n, 0 \longrightarrow \mathbb{C}^p, 0$, with $f_0 \ \mathring{\pitchfork}_{alg} \ A$. Then, $A_0 = f_0^{-1}(A)$ is an almost free arrangement based on A and will be called an A-*generic nonlinear arrangement.*

This corresponds in the linear case to (essential) A-generic arrangements.

By Theorem 4.1 and proposition 4.3, we have formulas for the singular Milnor number

$$(6.3) \quad \nu_A(f_0) = \dim_{\mathbb{C}} \mathcal{O}_{\mathbb{C}^s,0}\left\{\frac{\partial}{\partial y_i}\right\} / \left(\mathcal{O}_{\mathbb{C}^s,0}\left\{\frac{\partial f_0}{\partial x_1}, \dots, \frac{\partial f_0}{\partial x_n}, \zeta_1 \circ f_0, \dots, \zeta_{p-1} \circ f_0\right\}\right)$$

and higher multiplicities

(6.4) $$\mu_k(A_0) = \nu_A(f_0 \circ i).$$

where i is an embedding of a k-plane and $\nu_A(f_0 \circ i)$ is minimum among nearby embeddings i'.

Then, as in the linear case, we would like in the nonlinear weighted homogeneous case to obtain from (6.3) and (6.4) numerical formulas based on exp(A) and the weights of f_0. For an arrangement other than a Boolean one, not all $e_i = 1$; hence, even if the coordinate functions of f_0 are homogeneous the $\zeta_j \circ f_0$ need not be weighted homogeneous. It is still possible to use Theorem 2 of [D4] provided the generators in (6.3) are semi-weighted homogeneous in the sense of corollary 1.5 of [D4].

However, we concentrate on cases where the generators in (6.5) are weighted homogeneous.

(6.5) $$\left\{\frac{\partial f_0}{\partial x_1}, \dots, \frac{\partial f_0}{\partial x_n}, \zeta_1 \circ f_0, \dots, \zeta_{p-1} \circ f_0\right\}$$

This requires that we can assign weights $wt(x_i) = a_i$, and $wt(y_j) = c_j$, so that for weighted homogeneous $f_0 = (f_{01}, \dots, f_{0p})$, with f_{0j} of weighted degree d_j, $\zeta_{ji} \circ f_0$ is weighted homogeneous for these weights so

$$wt\left(\frac{\partial f_{0j}}{\partial x_i}\right) = d_j - a_i \qquad \text{and} \qquad wt(\zeta_{ji} \circ f_0) = d_i + c_j$$

Then, the matrix of degrees for the generators in (6.5) is given by (6.6).

(6.6) $$\mathcal{D}(A, f_0) = \begin{pmatrix} d_1-a_1 & d_2-a_1 & d_3-a_1 & \dots & d_p-a_1 \\ d_1-a_2 & d_2-a_2 & d_3-a_2 & \dots & d_p-a_2 \\ \cdots & \cdots & \cdots & \dots & \cdots \\ d_1-a_n & d_2-a_n & d_3-a_n & \dots & d_p-a_n \\ e_1 d_1 & e_1 d_2 & \cdots & \dots & e_1 d_p \\ \cdots & \cdots & \cdots & \dots & \cdots \\ e_{p-1} d_1 & e_{p-1} d_2 & \cdots & \dots & e_{p-1} d_p \end{pmatrix} \begin{matrix} \left.\vphantom{\begin{matrix}a\\a\\a\\a\end{matrix}}\right\} n \\ \left.\vphantom{\begin{matrix}a\\a\\a\end{matrix}}\right\} p-1 \end{matrix}$$

Remark : In the semi-weighted homogeneous case we would use instead the initial parts of the generators in (6.5)for the same weight conditions.

We determine from (6.6) the associated degree matrix (6.8) for $n \geq p$.

Remark 6.7: If $n < p$, then we instead obtain $\mathbf{D}_A(f_0)$ as the matrix formed from the last n rows of the matrix in (6.8).

$$(6.8) \quad D_A(f_0) \;=\; \begin{pmatrix} d_1-a_1 & d_2-a_2 & d_3-a_3 & \cdots & \cdot\cdot & d_p-a_p \\ d_1-a_2 & d_2-a_3 & d_2-a_4 & \cdots & \cdot\cdot & \cdot\cdot \\ d_1-a_3 & d_2-a_4 & d_3-a_5 & \cdots & \cdot\cdot & d_p-a_n \\ d_1-a_4 & d_2-a_5 & \cdot\cdot & \cdots & d_{p-1}-a_n & e_1d_p \\ \cdot\cdot & \cdot\cdot & \cdot\cdot & \cdots & e_2d_{p-1} & e_2d_p \\ \cdot\cdot & \cdot\cdot & \cdot\cdot & \cdots & \cdot\cdot & \cdot\cdot \\ d_1-a_n & e_{p-1}d_2 & \cdot\cdot & & \cdots & e_{p-1}d_p \end{pmatrix} \begin{matrix} \left.\begin{matrix} \\ \\ \\ \end{matrix}\right\} n-p+1 \\ \left.\begin{matrix} \\ \\ \\ \\ \end{matrix}\right\} p-1 \end{matrix}$$

Then, by Theorem 2 of [D4], we can compute the singular Milnor number $\nu_A(f_0)$ using $\tau(\mathbf{D}_A(f_0))$. We then will evaluate this in a number of situations.

Theorem 6.9 : *Provided the generators in (6.5) are (semi) weighted homogeneous, for the weights in (6.6), then the nonlinear arrangement defined by* f_0 *has its singular Milnor number given by*

$$\nu_A(f_0) \;=\; (1/a)\cdot\tau(\mathbf{D}_A(f_0))$$

where $a = \prod_{i=1}^{n} a_i$.

Homogeneous Nonlinear Arrangements:

We consider two key situations when we can directly apply Theorem 6.9: i) an arrangment of hypersurfaces of fixed degree d+1 based on a free arrangement A and ii) the case of a Boolean arrangement of homogeneous hypersurfaces of varying degrees d_i.

As in the linear case, we refer to a nonlinear arrangement based on a Boolean arrangement as a *generic nonlinear arrangement.* This is a union of isolated hypersurface singularities in algebraic general position off 0.

To apply Theorem 6.9, we use results from [D4]. For these, in addition to the elementary symmetric functions $\sigma_k(\mathbf{x})$ we also use $s_k(\mathbf{x})$, which is the polynomial defined as the sum of all monomials of degree k in $\mathbf{x} = (x_1, \dots, x_p)$.

Proposition 6.10 : *For a generic nonlinear arrangement* A_0 *of* p *homogeneous hypersurfaces* of degrees d_j, *the singular Milnor number is given by*

$$(6.11)\quad \mu(A_0) = s_n(\mathbf{d-1}) + \binom{p-1}{1}\cdot s_{n-1}(\mathbf{d-1}) + \dots + \binom{p-1}{n-1}\cdot s_1(\mathbf{d-1}) + \binom{p-1}{n}$$

where $\binom{\ell}{q} = 0$ *if* $q > \ell$.

Also, the higher multiplicities $\mu_j(A_0)$ *are obtained from* (6.11) *by replacing* n *by* j.

Proof : For the Boolean arrangement all $e_i = 1$. Hence, we can assign weights $wt(x_i) = 1$, and $wt(y_j) = d_j = \deg(f_{0j})$. Then, Theorem 6.9, applies. In the notation of (2.22) of [D4], $\mathbf{D}_A(f_0) = \mathbf{D}_n(p-1, \mathbf{d-1})|_{x=1}$ with $\mathbf{d-1} = (d_1-1, \dots, d_p-1)$ (and $k = p$). To compute $\tau(\mathbf{D}_A(f_0))$, we apply corollary 2.23 of [D4]. □

Second, we consider a nonlinear arrangement of hypersurfaces of fixed degree d+1, but based on any free arrangement A.

Proposition 6.12 : *For an* A*-generic nonlinear arrangement* A_0 *of hypersurfaces each of degree* d+1, with $\exp(A) = (e_0, \dots, e_{p-1})$ *the singular Milnor number*

$$\mu(A_0) = \sigma_n(d^n, e_1\cdot(d+1), \dots, e_{p-1}\cdot(d+1));$$

and the higher multiplicities are given by

$$\mu_k(A_0) = \sigma_k(d^k, e_1\cdot(d+1), \dots, e_{p-1}\cdot(d+1)).$$

Proof : Since all $d_i = d+1$, the generators in (6.5) are homogeneous of degrees $(d^n, e_1\cdot(d+1), \dots, e_{p-1}\cdot(d+1))$, i.e. $(d, d, \dots, d, e_1\cdot(d+1), \dots, e_{p-1}\cdot(d+1))$ with n factors of d for μ. Thus, theorem 1 of [D4] implies the result. For μ_k it is similar except there will be k factors of d. □

These two results agree for a generic arrangements of hypersurfaces of degree d+1, for which all $e_i = 1$.

Corollary 6.13 : *For a generic arrangement* A_0 of p *hypersurfaces each of degree* d+1,
the singular Milnor number $\mu(A_0) = \sigma_n(d^n, (d+1)^{p-1})$;
and the higher multiplicities are given by

$$\mu_k(A_0) = \sigma_k(d^k, (d+1)^{p-1}).$$

Examples 6.1 revisited: We can apply the preceding results to the examples in (6.1). Examples a) and b) are both singular Milnor fibers for nonlinear generic arrangements denoted A_a or A_b. The number of singular vanishing cycles are given by corollary 6.13:

a) $\mu(A_a) = \sigma_2(1^2, 2)\ (= \sigma_2(1, 1, 2)) = 1+2+2 = 5;$

b) $\mu(A_b) = \sigma_2(1^2, 2^2)\ (= \sigma_2(1, 1, 2, 2)) = 1\cdot 1 + 2\cdot 2 + 4(1\cdot 2) = 13;$

Example c) is the singular Milnor fiber for nonlinear generic arrangement A_c of three hyperplanes and a quadric so $\mathbf{d} = (1, 1, 1, 2)$.
Then, by proposition 6.10

$$\mu(A_c) = s_2(0^3, 1) + \binom{3}{1} s_1(0^3, 1) + \binom{3}{2} = 1+3+3 = 7$$

Alternately, $\mu(A_c) = \tau(\mathbf{D}) = 7$, where by remark (6.7) (since $n = 2 < 4 = p$) $\mathbf{D}$ is given by

$$\mathbf{D} = \begin{pmatrix} 0 & 0 & 1 & 2 \\ 0 & 1 & 1 & 2 \end{pmatrix}$$

Remark : In these cases, we are able to realize the singular vanishing cycles via real arrangements much as for the case of hyperplanes. This is not always possible for nonlinear arrangements.

Weighted Homogeneous Generic Arrangements

For a generic arrangement of p weighted homogeneous hypersurfaces of weighted degrees d_j (with $wt(x_i) = a_i$), the singular Milnor number is given by the following.

Proposition 6.14 : *For a nonlinear generic arrangement* A_0 *of weighted homogeneous hypersurfaces of weighted degrees* d_j, *the singular Milnor number* $\mu(A_0)$ is given by

$$(1/a)\cdot\Big((-1)^n\cdot\sigma_n(\mathbf{a}) + (-1)^{n-1}\cdot\sigma_{n-1}(\mathbf{a})\cdot s_1(\mathbf{d}) + \dots + (-1)\sigma_1(\mathbf{a})\cdot s_{n-1}(\mathbf{d}) + s_n(\mathbf{d})\Big)$$

where $a = \prod a_i$

Remark : In the homogeneous case, all $a_i = 1$ so $\sigma_k(\mathbf{a}) = \binom{n}{k}$. Then, the answer given by (6.14) looks quite different from (6.11); however, they are equivalent.

Proof : All $e_i = 1$ so we see from (6.6) that the generators in (6.5) are weighted homogeneous with weights ($\mathbf{b}$: $\mathbf{a}$, $\mathbf{d}$) where $\mathbf{b} = (-a_1, \dots, -a_n, 0, \dots, 0)$. Thus, in the notation of (2.18) of [D4], the degree matrix $\mathbf{D}_A(f_0)$ has the form $\mathbf{D}(\mathbf{b}, \mathbf{d})$. We can compute $\tau(\mathbf{D}_A(f_0))$ by applying results of [D4], either Corollary 3 or Theorem 2 together with proposition 2.19 (of [D4]) to obtain the result. For this we evaluate

$$\sigma_k(\mathbf{b}) \;=\; \sigma_k(-a_1, \dots, -a_n, 0, \dots, 0) \;=\; (-1)^k\cdot\sigma_k(a_1, \dots, a_n). \qquad \square$$

Transverse Union of a Linear Arrangement and Isolated Hypersurface Singularity :

Let $A' \subset \mathbb{C}^n$ be an almost free arrangement based on $A \subset \mathbb{C}^p$ (via φ) and let $f_0 : \mathbb{C}^n,0 \to \mathbb{C},0$ define an isolated hypersurface singularity V which is algebraically transverse to A′ off 0. Then, A″ = A′ ∪ V is an almost free nonlinear arrangement based on $A \uplus \{0\} \subset \mathbb{C}^{p+1}$. For example, c) of example (6.1) has this form. As we shall see in §8, the properties of this nonlinear arrangement are related to those of the restriction of the arrangement A′ to the Milnor fiber of f_0. Then, the singular Milnor number of A″ is computed by the following.

Proposition 6.15 : i) *If the hypersurface singularity* V *is homogeneous of degree* d+1 *then the singular Milnor number of* A″ *is given by the polynomial*

$$\mathcal{P}_n(\mathbf{e})(d) = d^n + \sigma_1(\mathbf{e})\cdot d^{n-1} + \ldots + \sigma_{n-1}(\mathbf{e})\cdot d + \sigma_n(\mathbf{e})$$

for $\exp(A) = \mathbf{e} = (e_0, \ldots, e_{p-1})$ where again $\sigma_k(\mathbf{e})$ is the k-th elementary symmetric function.

ii) *In the case that* A′ = A *is itself free, but without* V *being homogeneous, then the singular Milnor number of* A″ *is given by*

$$\dim_{\mathbb{C}}\left(\mathcal{O}_{\mathbb{C}^n,0}/(f_0+\zeta_0(f_0), \zeta_1(f_0), \ldots, \zeta_{p-1}(f_0))\right)$$

where $\{\zeta_0, \zeta_1, \ldots, \zeta_{p-1}\}$ *generate* Derlog(A), *and* ζ_0 *is the Euler vector field.*

Proof : Both of these results follow as special cases of the general calculation.

Suppose A is defined by the homogeneous equation H so that $\{\zeta_1, \zeta_2, \ldots, \zeta_{p-1}\}$ generate Derlog(H) and ζ_0 denotes the Euler vector field on $\mathbb{C}^p$. Then, if z denotes a coordinate for $\mathbb{C}$, $\{\zeta_0 - z\cdot\frac{\partial}{\partial z}, \zeta_1, \zeta_2, \ldots, \zeta_{p-1}\}$ generate Derlog(zH).

Second, we may choose local coordinates so that φ is the inclusion $\varphi(x) = (x, 0)$. Then, A″ is defined from $B = A \uplus \{0\}$ via $\psi = (\varphi, f_0): \mathbb{C}^n,0 \longrightarrow \mathbb{C}^{p+1},0$. We obtain via Theorem 4.1, that $\nu_B(\psi)$ is given by

$$\text{(6.16)} \quad \dim_{\mathbb{C}} (\mathcal{O}_{\mathbb{C}^n,0})^{p+1}/\mathcal{O}_{\mathbb{C}^n,0}\{(\zeta_0 - z\cdot\frac{\partial}{\partial z})\circ\psi, \zeta_1\circ\psi, \ldots, \zeta_{p-1}\circ\psi, \frac{\partial\psi}{\partial x_1}, \ldots, \frac{\partial\psi}{\partial x_n}\}$$

Since

$$\frac{\partial\psi}{\partial x_i} = \frac{\partial}{\partial x_i} + \frac{\partial f_0}{\partial x_i},$$

we can project onto the last p−n+1 factors. If we write $\zeta_i = \zeta_i'' + \zeta_i'$ (the $\mathbb{C}^n$ and $\mathbb{C}^{p-n}$ components) and use $\zeta_i\circ\varphi = \zeta_i\circ\psi = \zeta_i \mid \mathbb{C}^n$, then (6.16) equals

$$\text{(6.17)} \quad \dim_{\mathbb{C}}(\mathcal{O}_{\mathbb{C}^n,0})^{p-n+1}/M \quad \text{where}$$

$$M = \mathcal{O}_{\mathbb{C}^n,0}\{(\zeta_0' - f_0 - \zeta_0''(f_0)), (\zeta_1' - \zeta_1''(f_0)), \ldots, (\zeta_{p-1}' - \zeta_{p-1}''(f_0))\}$$

In case ii) $n = p$, $\varphi = \text{id}$ and all $\zeta_i{}' = 0$; thus we obtain exactly the formula for ii).

In case i), we see that if f_0 is homogeneous of degree $d+1$, then the generators of the submodule M in (6.17) are weighted homogeneous where we use weights $wt(x_i) = 1$ all i, $wt(y_j) = 1$ for $j \geq n$, and $wt(z) = d+1$. The weights of the generators of M are then exactly $e_0, \ldots, e_{p-1}$. The degree matrix is (6.18) which is exactly the matrix $P_n(d, x)$ of (2.21) of [D4] with $\mathbf{d} = \mathbf{e}$ and $x = d$. Thus, by corollary 2.21 of [D4], case i) follows. □

$$\textbf{(6.18)} \qquad P_n(\mathbf{e}, d) = \begin{pmatrix} e_0 & e_1 & e_2 & \cdots\cdots & e_{p-n-2} & e_{p-n-1}+d \\ e_1 & e_2 & e_3 & \cdots & e_{p-n-1} & e_{p-n}+d \\ e_2 & e_3 & \cdot\cdot & & \cdots & \cdot\cdot \\ \cdot\cdot & \cdot\cdot & \cdot\cdot & \cdots & & \cdot\cdot \\ e_{n-1} & e_n & e_{n+1} & & \cdots & e_{p-1}+d \end{pmatrix}$$

Remark 6.19 : Observe that (6.17) actually gives a general answer in all cases provided it can be explicitly computed.

Part III Almost Free Complete Intersections

We extend the notion of almost free divisor to the case of complete intersections (which need not be isolated). In §7, we deduce the main properties : almost free complete intersections are closed under transverse intersection and pull-back by algebraically transverse finite map germs; and transverse intersections of almost free divisors yield almost free complete intersections. By applying arguments in [DM] which use a theorem of Lê (recall §4), we can associate both singular Milnor fibers and numbers and higher multiplicities to these complete intersections. By earlier transversality results, the higher multiplicities can be also be computed as singular Milnor numbers of almost free complete intersections. In preparation for §8, we give the formula for the singular Milnor number of a transverse union of almost free divisors.

In §8, we obtain formulas for singular Milnor numbers (theorem 2) and higher multiplicities (corollary 8.9) of almost free complete intersections as alternating sums of singular Milnor numbers for associated nonlinear arrangements. In the weighted homogeneous case, this gives formulas in terms of the weighted Macaulay-Bezout numbers and the function τ. By applying the formulas for nonlinear arrangements obtained in §6, we obtain explicit formulas for a number of situations, including: ICIS (recovering formulas of Greuel-Hamm and Giusti), almost free arrangements on ICIS, nonlinear mappings of discriminants, intersections of discriminants, etc.

In § 9 we give an alternate method to compute singular Milnor numbers by proving an analogue of the Lê-Greuel formula for almost free complete intersections. This formula is again given in terms of the length of a determinantal module.

§7 Singular Milnor Fibers for Complete Intersections

We begin by defining the notion of almost free complete intersection (which need not be isolated). First, we derive several basic properties (prop. 7.6) including: i) the class of almost free complete intersections is closed under transverse intersection and pull-backs by algebraically transverse finite map germs; and ii) the transverse intersection of almost free divisors is an almost free complete intersection. Second, we describe the stabilization of such germs giving rise to singular Milnor fibers and numbers. Third, we carry out a calculation of the singular Milnor number for transverse unions of almost free hypersurfaces needed for the next section.

Let $V_i' \subset \mathbb{C}^{p_i}$, $1 \leq i \leq k$, be free divisors with good defining equations $H_i : \mathbb{C}^{p_i},0 \longrightarrow \mathbb{C},0$, and let $p = \sum p_i$.

Definition 7.1 : An *almost free complete intersection* V based on the free divisors $V_i' \subset \mathbb{C}^{p_i}$, $1 \leq i \leq k$, is defined via a germ $f_0: \mathbb{C}^n,0 \longrightarrow \mathbb{C}^p,0$ with

$f_0 \mathring{\pitchfork}_{alg} \prod V_i'$ by $V = f_0^{-1}(\prod V_i')$. If we use instead geometric transversality we obtain a "*geometrically almost free complete intersection*".

Comments for geometric version:.

First, the Whitney stratification we use is obtained by taking the canonical Whitney stratification $\{S_i^{(j)}\}$ of V_j' and adding $S_0^{(j)} = \mathbb{C}^{p_j} \setminus V_j'$. Then the products $\prod S_{i_j}^{(j)}$ are the strata of a Whitney stratification for $\mathbb{C}^p$. These will also give a Whitney statification of all of the subspaces of $\mathbb{C}^p$ we will consider ,

such as $\prod V_i'$, $\circledast\ V_i'$, etc.

Second, *all of the results in this section and the next up through corollary 7.19, although stated for the algebraic case, will be valid for the geometric one.* For simplicity, we will state and prove all of the results for the algebraic case.

For example, the proofs for results using geometric transversality are analogous to the algebraic ones using $T_y S_i^{(j)}$ in place of $T_{\log}(V_j')_{(y)}$. However, we will refer to the "geometric versions" of the various results.

The next lemma extends several of the transversality results from part I, not requiring special properties of V_i'.

Lemma 7.2 : i) *Suppose* $V_i \subset \mathbb{C}^n$ *are given by* $\varphi_i^{-1}(V_i')$ *for* $\varphi_i : \mathbb{C}^n,0 \to \mathbb{C}^{p_i},0$ *with* $\varphi_i \,\mathring{\pitchfork}_{alg}\, V_i'$. *Then for* $\varphi = (\varphi_1, \dots, \varphi_k) : \mathbb{C}^n,0 \to \mathbb{C}^p,0$,

V_i *are in algebraic general position off* 0 *iff* $\varphi \,\mathring{\pitchfork}_{alg} \,\uplus\, V_i'$.

ii) *If as in* i), $\varphi \,\mathring{\pitchfork}_{alg} \,\uplus\, V_i'$, then $\varphi \,\mathring{\pitchfork}_{alg} \prod V_i'$.

iii) *More generally, if* $\varphi \,\mathring{\pitchfork}_{alg} \,\uplus\, V_i'$ *then* $\varphi_{\mathbf{j}} \,\mathring{\pitchfork}_{alg} \prod V_{j_i}'$ for $\mathbf{j} = \{j_1,\dots, j_r\} \subset \{1,\dots, k\}$ *and* $\varphi_{\mathbf{j}} = (\varphi_{j_1},\dots, \varphi_{j_r}) : \mathbb{C}^n,0 \to \mathbb{C}^{p_{\mathbf{j}}},0$ *with* $p_{\mathbf{j}} = \sum p_{j_i}$

iv) *The preceding remain valid for* $\varphi : U \to \mathbb{C}^p$ *where* $\varphi(U) \subset U'$ *and the* V_i' *denote representatives on* U'.

Proof : For i) we first apply ii) of proposition 3.1 but for k factors

$$T_{\log}(\uplus\, V_i')_{y_0} = \oplus\, T_{\log}(V_i')_{(y_{0i})} \tag{7.3}$$

where $y_0 = (y_{01}, \dots, y_{0k}) \in \mathbb{C}^p$. Thus, algebraic transversality to $\uplus\, V_i'$ is equivalent to the surjectivity of

$$df_0(x_0): T_{x_0}\mathbb{C}^n \longrightarrow \oplus\, T_{y_{0i}}\mathbb{C}^{p_i}/T_{\log}(V_i')_{(y_{0i})} \tag{7.4}$$

If $\mathbf{j} = \{ j : x_0 \in V_j\}$ then

$$T_{y_{0i}}\mathbb{C}^{p_i}/T_{\log}(V_i')_{(y_{0i})} = (0) \quad \text{if } i \notin \mathbf{j}.$$

Thus, (7.4), but only summed over $i \in \mathbf{j}$, is surjective iff

$$\{ df_0(x_0)^{-1}(T_{\log}(V_i')_{(y_{0i})}) : i \in \mathbf{j} \} \text{ are in general position.} \tag{7.5}$$

By (2.8)

$$df_0(x_0)^{-1}(T_{\log}(V_i')_{(y_{0i})}) = T_{\log}(V_i)_{(x_0)}.$$

Thus, the result again follows by standard linear algebra.

For ii, we observe that each $\zeta \in \mathrm{Derlog}(V_i')$ is, via a trivial lift to $\mathbb{C}^p$, also in $\mathrm{Derlog}(\prod V_i')$ (if $I(V_i')$is the defining ideal for V_i' , then $I(\prod V_i') = \prod I(V_i')$, then apply ζ). Thus, $\oplus \mathrm{Derlog}(V_i') \subseteq \mathrm{Derlog}(\prod V_i')$ so

$$\oplus T_{\log}(V_i')_{(y_{0i})} \subseteq T_{\log}(\prod V_i')_{(y_0)}.$$

Transversality follows from that for $\circledast\ V_i'$ by (7.3).

For iii), it is only necessary to observe that the transversality of φ implies (7.5) holds for any $\mathbf{j}' \subseteq \mathbf{j}$, which is exactly by i) the condition for

$\varphi_j \mathrel{\mathring{\pitchfork}_{alg}} \circledast\ V_{j_i}'$. By ii) this implies the desired transversality to $\prod V_{j_i}'$.

The preceding arguments are seen to also apply in case iv). □

We obtain the following properties of almost free complete intersections.

Proposition 7.6 : i) *Suppose* $V_i \subset \mathbb{C}^n$ *are almost free divisors based on free divisors* $V_i' \subset \mathbb{C}^{p_i}$ *(via* $\varphi_i : \mathbb{C}^n,0 \to \mathbb{C}^{p_i},0$*). If* V_i *are in algebraic general position off* 0 *then,* $V = \cap V_i$ *is an almost free complete intersection defined via* $f_0 = (\varphi_1, \dots, \varphi_k) : \mathbb{C}^n,0 \to \mathbb{C}^p,0$.

ii) *If* V_1 , $V_2 \subset \mathbb{C}^n,0$ *are almost free complete intersections which are algebraically transverse off* 0 *then* $V_1 \cap V_2$ *is an almost free complete intersection;*

iii) *If* $V \subset \mathbb{C}^p,0$ *is an almost free complete intersection and* $f_0 : \mathbb{C}^n,0 \to \mathbb{C}^p,0$ *is a finite map germ with* $f_0 \mathrel{\mathring{\pitchfork}_{alg}} V$ *then* $V_0 = f_0^{-1}(V)$ *is an almost free complete intersection.*

Remark 7.7 : We shall refer to almost free complete intersections which satisfy i) of (7.6) *as defined by transverse intersection* (off 0). In the case of isolated complete intersections V, it is well-known that they can be represented as the transverse intersection of isolated hypersurface singularities V_i. Thus, they are almost free divisors in the stronger sense. In fact the same argument implies that given a finite number of singular germs W_i geometrically transverse to V off 0, we can find the V_i so that all $V_{j_1} \cap \dots \cap V_{j_r}$ are transverse to the W_i off 0. Thus, if $\{W_1, \dots, W_m, V\}$ are in general position off 0 so are $\{W_1, \dots, W_m, V_1, \dots, V_k\}$.

Proof : For i) we use ii) of Lemma 7.2.

For ii), let $V_1, V_2 \subset \mathbb{C}^n,0$ be defined by $V_1 = f_0^{-1}(\prod V_i')$ and $V_2 = g_0^{-1}(\prod W_i')$ where V_i' and W_i' are free divisors and with $f_0: \mathbb{C}^n,0 \to \mathbb{C}^p,0$ and $g_0: \mathbb{C}^n,0 \to \mathbb{C}^{p'},0$. Then, by i) of lemma 7.2, $f = (f_0, g_0) : \mathbb{C}^n,0 \to \mathbb{C}^{p+p'},0$ is algebraically transverse to $(\prod V_i') \times (\prod W_i')$ iff $V_1 \mathring{\pitchfork}_{alg} V_2$. Then, by ii) of lemma 7.2, $f \mathring{\pitchfork}_{alg} (\prod V_i') \times (\prod W_i')$, and $V_1 \cap V_2 = f^{-1}((\prod V_i') \times (\prod W_i'))$.

For iii), we apply corollary 2.11. □

Stabilizations for Almost Free Complete Intersections

Suppose that V is defined as an almost free complete intersection based on the free divisors $V_i' \subset \mathbb{C}^{p_i}$ and defined by $f_0 : \mathbb{C}^n,0 \to \mathbb{C}^p,0$. Again we consider a (geometric) stabilization of f_0. This is a family of maps $f_t : U \to \mathbb{C}^p,0$ such that f_0 is algebraically (resp. geometrically) transverse to $\prod V_i'$ on $U\backslash\{0\}$ and for t sufficiently small $\neq 0$, f_t is transverse to $\prod V_i'$ on U (where, as we mentioned above, in the geometric case the stata of the Whitney stratification are products of the canonical Whitney strata for each V_i'). We let $V_t = f_t^{-1}(\prod V_i')$. Then we can apply the arguments in [DM] to obtain

Lemma 7.8 : *For* ε *and* $t > 0$ *sufficiently small,* $V_t \cap B_\varepsilon$ *is homotopy equivalent to a bouquet of spheres of real dimension* $n-p$.

Proof : We let $F : \mathbb{C}^n \times \mathbb{C}, 0 \to \mathbb{C}^p,0$ be defined by $F(x, t) = f_t(x)$ and let $\mathcal{V} = F^{-1}(\prod V_i')$. Then, by applying the arguments in [DM], $\pi : \mathcal{V}, 0 \to \mathbb{C}, 0$ has an isolated singularity in the sense of of Lê [Lê1]. Thus, applying a theorem of Lê for such isolated singularities [Lê1] [Lê2], we obtain the result. □

Remark 7.9 : Just as for the hypersurface case, stratification theory implies the space $V_t \cap B_\varepsilon$ is independent of the stabilization f_t and will be referred to as the *singular Milnor fiber* of the V. The number of spheres will also be called the *singular Milnor number* and denoted by $\mu_{V'}(f_0)$ (for V based on $V' = \prod V_i'$ via f_0) or just $\mu(V)$ when the representation of V is clear from the context.

Finally, we can also define the higher multiplicities. We again consider a generic section $i\colon P \hookrightarrow \mathbb{C}^n$ of V (for simplicity we assume $P = \mathbb{C}^s$) which is geometrically transverse to V off 0. Again let $i_t\colon \mathbb{C}^s,0 \hookrightarrow \mathbb{C}^n,0$ be a (topological) stabilization of i. By the geometric version of proposition 2.12, $i \circ f_0$ deines a geometrically almost free complete intersection. the singular Milnor fiber for $i = i_t(\mathbb{C}^s) \cap V \cap B_\varepsilon$, for B_ε a ball of radius $\varepsilon > 0$ and ε and t sufficiently small. We let $\mu_s(V)$ denote this singular Milnor number. We refer to $\{\mu_s(V)\}$ as the *set of higher multiplicities for* V (by definition, $\mu_0(V) = 1$).

Remark 7.10 : Also, if V is defined as the transverse intersection of $\cap V_i$ of almost free divisors via $f_0 = \varphi = (\varphi_1, \dots, \varphi_p) : \mathbb{C}^n,0 \to \mathbb{C}^p,0$, then its possible to choose a stabilization f_t for f_0 relative to $\times V_i'$. Then, by lemma 7.2, it is also a stabilization for any subset $\cap V_{j_i}$ which is also an almost free complete intersection.

Singular Milnor Numbers of Transverse Unions of Almost Free Divisors

In preparation for the computations in the next section of the singular Milnor numbers for almost free complete intersections, we explicitly give the formula for the transverse union of almost free divisors.

For $1 \leq i \leq k$, let $V_i \subset \mathbb{C}^n$ be an almost free divisor based on the free divisor $V_i' \subset \mathbb{C}^{p_i}$ via $\varphi_i : \mathbb{C}^n,0 \to \mathbb{C}^{p_i},0$. We suppose we have good defining equations $H_i : \mathbb{C}^{p_i},0 \longrightarrow \mathbb{C},0$, and let $p = \sum p_i$. Let $\{\zeta_1^{(j)}, \dots, \zeta_{p_j-1}^{(j)}\}$ denote the generators for Derlog(H_j) with $\zeta_0^{(j)}$ the "Euler vector field". We write

(7.11) $\{\zeta_1^{(1)}, \dots, \zeta_{p_1-1}^{(1)}, \zeta_1^{(2)}, \dots, \dots, \zeta_{p_k-1}^{(k)}, \zeta_0^{(1)}-\zeta_0^{(2)}, \dots, \zeta_0^{(1)}-\zeta_0^{(k)}\}$
in abbreviated form as $\{\zeta_i^{(j)}\ (i \geq 1), \zeta_0^{(1)}-\zeta_0^{(i)}\}$.

Then, the singular Milnor number for $\cup V_i$ is given by the following.

Proposition 7.12 : *If* $V_i \subset \mathbb{C}^n$ *are almost free divisors based on* V_i' *as above, for* $1 \leq i \leq k$, *such that* $n < \hbar(V_i')$ *for each* i, *then*

$$\mu(\cup V_i) = \nu_{V'}(f_0)$$

$$(7.13) \qquad = \dim_{\mathbb{C}}\left(\mathcal{O}_{\mathbb{C}^n,0}\right)^p / \left(\mathcal{O}_{\mathbb{C}^n,0}\left\{\frac{\partial f_0}{\partial x_i}, \zeta_i^{(j)}\circ f_0\ (i \geq 1), \zeta_0^{(1)}\circ f_0 - \zeta_0^{(i)}\circ f_0\right\}\right)$$

There is a similar formula for $\mu_s(\cup V_i) = \nu_{V'}(f_0 \circ i)$ *by replacing* $\mathbb{C}^n$ *in* (7.13) *by* $\mathbb{C}^s$ *and* f_0 *by* $f_0 \circ i$ *for a sufficiently general embedding* i.

Proof : If the V_i are in algebraic general position off 0, then by corollary 3.9 and i) of Lemma 7.2, $V = \cup V_i$ is an almost free divisor based on $V' \overset{\text{def}}{=} \times V_i'$ via

$$f_0 = (\varphi_1, \dots, \varphi_k) : \mathbb{C}^n,0 \longrightarrow \mathbb{C}^p,0.$$

To compute its singular Milnor numbers and higher multiplicities using the results of §4, we need to determine Derlog(H) for $H = \prod H_i$ the good defining equation for V′. Then, by repeatedly applying Remark 3.3, we see that (7.11) is a set of generators for Derlog(H).

Also, to apply Theorem 4.1 or Proposition 4.3, we need to know $n < \hbar(V')$. However, the strata for the Whitney stratification of V′ are of the form $\prod S_{i_j}^{(j)}$ for $S_i^{(j)}$ the Whitney strata of V_j' or $S_0^{(j)} = \mathbb{C}^{p_j} \setminus V_j'$. Also, if $y_0 = (y_{01}, \dots, y_{0k}) \in \mathbb{C}^p$, and $y_{0j}' = y_{0j}$ if $y_{0j} \in V_j'$ and $y_{0j}'' = y_{0j}$ if $y_{0j} \notin V_j'$. Then,

$$(7.14) \qquad \left(\oplus\, T_{\log}(H_i)_{(y_{0i}')}\right) \oplus \left(\oplus\, T\,\mathbb{C}^{p_j}_{(y_{0j}'')}\right) \subseteq T_{\log}(H)_{y_0}.$$

Hence,

$$\hbar(V') \geq \min\{\hbar(V_i')\}$$

Thus, provided $n < \hbar(V_i')$ for each i, then $n < \hbar(V')$. Hence, we can apply Theorem 4.1 and Proposition 4.3 to obtain the results. □

Example 7.15 : In the special case when all $V_i' = \{0\} \subset \mathbb{C}$, the generators in (7.11) become $\{ y_1 \cdot \frac{\partial}{\partial y_1} - y_2 \cdot \frac{\partial}{\partial y_2}, \dots, y_1 \cdot \frac{\partial}{\partial y_1} - y_p \cdot \frac{\partial}{\partial y_p} \}$. These are also the generators used for a Boolean arrangement in $\mathbb{C}^p$. This is as it should be since the transverse union of isolated hypersurface singularities is a generic arrangement.

We are now ready to compute the singular Milnor numbers for almost free complete intersections via the formulas for transverse unions which, in the special case of isolated complete intersections, reduces to the singular Milnor numbers for generic nonlinear arrangements.

§8 Computing Singular Milnor Numbers for Almost Free Complete Intersections

We shall give formulas for the singular Milnor numbers and higher multiplicities for almost free complete intersections represented as transverse intersections of almost free divisors. We apply these results to a number of situations including: ICIS (recovering the formulas of Greuel-Hamm and Giusti), isolated complete intersections on almost free divisors (or vice versa), arrangements on isolated complete intersections, and isolated hypersurface singularities on discriminants.

Remark : In all of the statements, we will state results for algebraically transversality; however, *all of the results up through corollary 8.9 plus proposition 8.17 are equally valid for geometric transversality and general position.*

Let $V_i \subset \mathbb{C}^n$, $1 \leq i \leq k$. For a subset $\mathbf{j} = \{j_1, \ldots, j_r\} \subseteq \{1, \ldots, k\}$, we let $V_{\cap \mathbf{j}} = V_{j_1} \cap \ldots \cap V_{j_r}$; and similarly for $V_{\cup \mathbf{j}}$ replacing $\cap$ by $\cup$, with $V_\cap$ and $V_\cup$ denoting the case $\mathbf{j} = \{1, \ldots, k\}$. Then, there is the following formula for the singular Milnor number of an almost free complete intersection in terms of those for almost free nonlinear arrangements.

Theorem 2 : *Let* $V_i \subset \mathbb{C}^n$, $1 \leq i \leq k$, *be almost free divisors which are in algebraic general position off* 0. *Then,*

i) $$\mu(V_\cap) = \sum (-1)^{|\mathbf{j}|+k} \mu(V_{\cup \mathbf{j}})$$

or equivalently

ii) $$\mu(V_\cup) = \sum \mu(V_{\cap \mathbf{j}})$$

(*where both sums are over nonempty* $\mathbf{j} \subseteq \{1, \ldots, k\}$).

Remark : It should be remarked that despite its unexpected form, the theorem can be used very effectively to compute singular Milnor numbers. This is because each $V_{\cup \mathbf{j}}$ is an almost free divisor, so we can compute the singular Milnor numbers and higher multiplicities using e.g. proposition 7.12. In turn, these can be computed in the weighted homogeneous case as Macaulay-Bezout numbers. Moreover, the formulas for each $V_{\cup \mathbf{j}}$ often have the form of a linear combination $\sum \alpha_{\mathbf{j}i}\psi_i$ of fixed terms ψ_i independent of $\mathbf{j}$, with coefficients $\alpha_{\mathbf{j}i}$ depending on $\mathbf{j}$. Thus, for the sum it is only necessary to find an expression for the sum of the coefficients.

Proof : We shall see that with the results already obtained, the proof reduces to simple Euler characteristic arguments. We begin with a few simple facts about Euler characteristics. Suppose we are given subcomplexes of a CW-complex $A_i \subseteq X$, $1 \le i \le k$. For a subset $\mathbf{j} = \{j_1, \dots, j_r\} \subseteq \{1, \dots, k\}$, we let $A_{\cap \mathbf{j}} = A_{j_1} \cap \dots \cap A_{j_r}$; and similarly for $A_{\cup \mathbf{j}}$ replacing $\cap$ by $\cup$, with $A_\cap$ and $A_\cup$ denoting the case $\mathbf{j} = \{1, \dots, k\}$. Each of the $A_{\cap \mathbf{j}}$ is also a subcomplex. Then, there are the following formulas for $\chi(A_\cap)$ and $\chi(A_\cup)$.

Lemma 8.1 : i) $\chi(A_\cap) = \sum (-1)^{|\mathbf{j}|+1} \chi(A_{\cup \mathbf{j}})$

ii) $\chi(A_\cup) = \sum (-1)^{|\mathbf{j}|+1} \chi(A_{\cap \mathbf{j}})$

(summed over nonempty $\mathbf{j} \subseteq \{1, \dots, k\}$).

Proof : The proof involves nothing more than inductively applying Mayer-Vietoris and keeping track of signs. For example, for i) the proof is by induction on k. For $k = 1$ it is trivial, while for $k = 2$, Mayer-Vietoris implies

$$\text{(8.2)} \qquad \chi(A_1 \cap A_2) = \chi(A_1) + \chi(A_2) - \chi(A_1 \cup A_2).$$

If the result holds for $n < k$ then by (8.2)

$$\chi(A_1 \cap A_2 \cap \dots \cap A_k) = \chi(A_1) + \chi(A_2 \cap \dots \cap A_k) - \chi(A_1 \cup (A_2 \cap \dots \cap A_k))$$

By the induction hypothesis applied to the second and third terms on the RHS, we obtain

$$\chi(A_\cap) = \chi(A_1) + \sum (-1)^{|\mathbf{j}|+1} \chi(A_{\cup \mathbf{j}}) - \sum (-1)^{|\mathbf{j}'|+1} \chi(\cup(A_1 \cup A_{j'_i}))$$

where the sums are over nonempty $\mathbf{j}, \mathbf{j}' \subseteq \{2, \dots, k\}$. Bringing the minus sign inside in the last term, it can be written

$$\sum (-1)^{|\mathbf{j}'|+2} \chi(A_1 \cup A_{j'_1} \dots \cup A_{j'_m}) = \sum (-1)^{|\mathbf{j}''|+1} \chi(A_{\cup \mathbf{j}''})$$

for $\mathbf{j}'' = \{1\} \cup \mathbf{j}' = \{1, j'_1, \dots, j'_m\}$. Thus, we obtain the desired formula. □

Using this we can prove the theorem.

Proof (of the theorem) : Suppose the almost free divisors $V_i \subset \mathbb{C}^n$, $1 \leq i \leq k$, are based on free divisors $V_i' \subset \mathbb{C}^{p_i}$ via $\varphi_i : \mathbb{C}^n,0 \to \mathbb{C}^{p_i},0$. By assumption the V_i are in algebraic general position off 0. We consider a stabilization $f_t : U \to \mathbb{C}^p,0$ of $f_0 = (\varphi_1, \dots, \varphi_k)$ relative to the product union $\uplus\, V_i'$. Then, for any $j = \{j_1, \dots, j_r\} \subseteq \{1, \dots, k\}$, we can define $f_{jt} : U \to \mathbb{C}^{p_j},0$ where $p_j = \sum p_{j_i}$ and $f_{jt}(x) = (f_{j_1t}, \dots, f_{j_rt})$. By lemma 7.2, f_{jt} is transverse to $\uplus\, V_{j_i}'$ and provides a stabilization of f_{j0}.

We denote $V_{it} = f_{it}^{-1}(V_i')$. Then, the V_{it} are in (geometric) general position so for t sufficiently small and $\varepsilon > 0$ sufficiently small, iv) of lemma 7.2 implies $V_{\cap jt} \cap B_\varepsilon$ is a singular Milnor fiber for $V_{\cap j}$ where $V_{\cap jt} = \cap V_{j_it}$ and similarly $V_{\cup jt} \cap B_\varepsilon$ is a singular Milnor fiber for $V_{\cup j}$ where $V_{\cup jt} = \cup V_{j_it}$. Then, by lemma 7.8

(8.3) $$\chi(V_{\cap jt} \cap B_\varepsilon) = 1 + (-1)^{n-|j|}\mu(V_{\cap j})$$

and $$\chi(V_{\cup jt} \cap B_\varepsilon) = 1 + (-1)^{n-1}\mu(V_{\cup j}).$$

We prove i) with a similar argument for ii). By i) of lemma 8.1 applied to $A_j = V_{jt} \cap B_\varepsilon$ and (8.3) we obtain

(8.4) $$1 + (-1)^{n-k}\mu(V_\cap) = \sum (-1)^{|j|+1} (1 + (-1)^{n-1}\mu(V_{\cup j}))$$

For the terms involving 1 on the RHS, there are $\binom{k}{\ell}$ different j with $|j| = \ell$. Thus,

$$\sum (-1)^{|j|+1} = \sum_{\ell=1}^{k} (-1)^{\ell+1} \binom{k}{\ell} = 1.$$

Thus, from (8.4)

(8.5) $$(-1)^{n-k}\mu(V_\cap) = \sum (-1)^{|j|+1} (-1)^{n-1}\mu(V_{\cup j})$$

Multiplying (8.5) by $(-1)^{n-k}$ gives the result

$$\mu(V_\cap) = \sum (-1)^{|j|+k} \mu(V_{\cup j})$$ □

For example, for k = 2 this becomes

(8.6) $$\mu(V_1 \cap V_2) = \mu(V_1 \cup V_2) - \mu(V_1) - \mu(V_2)$$

A Simple Form of Duality for Intersections of Free Divisors :

If V_1 and V_2 are free divisors transverse off 0, then (8.6) yields

(8.7) $$\mu(V_1 \cap V_2) = \mu(V_1 \cup V_2)$$

This equality is between cycles of different dimensions. For example, if two free arrangements A_1 and A_2 intersect transversally off 0, then $A_1 \cup A_2$ is still an almost free arrangement by proposition 3.6. However, $A_1 \cap A_2$ is a union of codimension 2 planes so it is not even an arrangement of hyperplanes . Nonetheless, each has the same number of vanishing cycles.

In the case of the intersection of a hyperplane H with the Boolean arrangement A_3, we perturb H to H_t transverse to A_3. We see that both $H_t \cap A_3$ and $H_t \cup A_3$ each have one singular cycle represented by the triangle of intersection in H_t, respectively the (boundary of the) tetrahedron.

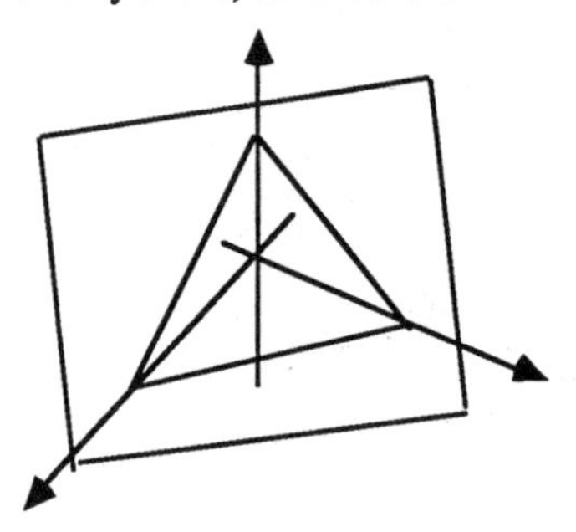

We can see the geometric correspondence giving the equality. The Mayer Vietoris sequence gives an isomorphism

$$H^{n-1}(V_{1t} \cap V_{2t}) \simeq H^n(V_{1t} \cup V_{2t})$$

If σ is a cycle in $V_{1t} \cap V_{2t}$ then $\sigma = \partial\sigma_1$ in V_{1t} ($\simeq V_1$), and similarly $\sigma = \partial\sigma_2$ in V_{2t} ($\simeq V_2$). Then, $\sigma_1 - \sigma_2$ is a cycle in $V_{1t} \cup V_{2t}$. This induces the bijection.

Suppose W is an almost free complete intersection and $V_i \subset \mathbb{C}^n$, $1 \leq i \leq k$, are almost free divisors which together with W are in algebraic general position off 0. Then, in the notation of theorem 2, $W \cap V_{\cup}$ is also an almost free complete intersection. By directly applying the arguments of theorem 2 we obtain as a corollary

Corollary 8.8 : *With* W *an almost free complete intersection and* $V_i \subset \mathbb{C}^n$, $1 \leq i \leq k$, *almost free divisors which together are in algebraic general position off* 0,

i) $$\mu(W \cap V_{\cup}) = \sum \mu(W \cap V_{\cap j})$$

or equivalently

ii) $$\mu(W \cap V_{\cap}) = \sum (-1)^{|j|+k} \mu(W \cap V_{\cup j})$$

(*where both sums are over nonempty* $\mathbf{j} \subseteq \{1, \dots, k\}$).

Remark : Here $W \cap V_{\cup}$ is an example of a nonlinear arrangement on the singular space W; moreover, its singular Milnor fiber can be obtained as the intersection of $V_{\cup}$ with the singular Milnor fiber of W. Thus, this is also a nonlinear arrangement. We shall be applying this to such arrangements shortly.

If we think of $\mu(W \cap V)$ as a higher order intersection number, then i) of (8.8) states that for transverse intersections off 0, the intersection number of W with $\cup V_i$ is the sum of all of the intersection numbers of W with all of the "obvious subvarieties" of $\cup V_i$.

By instead composing with a generic inclusion $i : \Pi \hookrightarrow \mathbb{C}^n$, we obtain as corollaries

Corollary 8.9 : *Let* $V_i \subset \mathbb{C}^n$, $1 \leq i \leq k$, *be almost free divisors which are in algebraic general position off* 0. *Then, the higher multiplicities are given by*

i) $$\mu_{\ell}(V_{\cup}) = \sum \mu_{\ell}(V_{\cap j})$$

ii) $$\mu_{\ell}(V_{\cap}) = \sum (-1)^{|j|+k} \mu_{\ell}(V_{\cup j})$$

(*where both sums are over nonempty* $\mathbf{j} \subseteq \{1, \dots, k\}$).

We apply these results to several different situations.

Isolated Complete Intersection Singularities :

We have already mentioned that isolated complete intersection singularities can be represented as the intersection of isolated hypersurface singularities and hence are almost free. We consider the weighted homogeneous case so $V \subset \mathbb{C}^n$ is defined by a weighted homogeneous germ $f = (f_1, \dots, f_p) : \mathbb{C}^n,0 \to \mathbb{C}^p,0$ where $wt(x_i) = a_i$ and $wt(y_j) = d_j$ with $wt(f_j) = d_j$. A formula for the Milnor number (and other invariants) in terms of the weights have been given by Greuel and Hamm [GrH] and Giusti [Gi]. We deduce this formula as a consequence of Theorem 2 and corollary 6.14.

We suppose that $V = \cap V_i$ is the transverse intersection of isolated hypersurface singularities where $V_i = f_i^{-1}(0)$. then $V_{\cup \mathbf{j}}$ is a nonlinear generic arrangement, hence by the results of §6 it can be computed by the function τ.

Theorem 8.10 (Greuel-Hamm, Giusti) : *For the isolated complete intersection* $V = \cap V_i$ *of p weighted homogeneous hypersurfaces of weighted degrees* d_i, *the Milnor number equals*

$$\mu(V) = (-1)^{n-p+1} + (d/a)\cdot\left(\sum_{j=0}^{n-p} (-1)^{n-p-j}\sigma_{n-p-j}(\mathbf{a})\cdot s_j(\mathbf{d}) \right)$$

where $a = \prod_{i=1}^{n} a_i$, $d = \prod_{i=1}^{p} d_i$ $(= \sigma_p(\mathbf{d}))$ (and $\sigma_0(\mathbf{d}) = s_0(\mathbf{c}) = 1$).

Remark : The sum in the formula can be expressed as $\tau(\mathbf{D}(-\mathbf{a}, (0,\mathbf{d})))$ in the notation of (2.18) of [D4]; however, at this time there is no obvious direct way of seeing it. Also, the derivation of this formula we give here is not as general as that of Giusti nor Greuel-Hamm in that we use a representation as a transverse intersection of weighted homogeneous isolated hypersurface singularities. This can always be achieved e.g. in the homogeneous case, but not in general.

Proof : By iii) of lemma 7.2, $V_{\cup \mathbf{j}}$ is a generic nonlinear arrangement of weighted homogeneous hypersurfaces for any nonempty subset $\mathbf{j} = \{j_1, \dots, j_r\} \subseteq \{1, \dots, p\}$. By proposition 6.14, the singular Milnor number $\mu(V_{\cup \mathbf{j}})$ equals

$$(1/a)\cdot\Big((-1)^n\cdot\sigma_n(\mathbf{a}) + (-1)^{n-1}\cdot\sigma_{n-1}(\mathbf{a})\cdot s_1(\mathbf{d}_\mathbf{j}) + \dots + (-1)\sigma_1(\mathbf{a})\cdot s_{n-1}(\mathbf{d}_\mathbf{j}) + s_n(\mathbf{d}_\mathbf{j})\Big)$$

where $\mathbf{d_j} = (d_{j_1}, \dots, d_{j_r})$. We observe that this expression is $(1/a)$ times a linear combination of $\sigma_k(\mathbf{a})$ whose coefficients only involve $s_i(\mathbf{d_j})$. Thus, when we apply Theorem 2 we will sum $(-1)^{p+|j|}$ times such expressions over nonempty $\mathbf{j} = \{j_1, \dots, j_r\} \subseteq \{1, \dots, p\}$. Thus, we obtain the answer as another linear combination

$$(1/a)\cdot\Big((-1)^n\gamma_0(\mathbf{d})\cdot\sigma_n(\mathbf{a}) + (-1)^{n-1}\sigma_{n-1}(\mathbf{a})\cdot\gamma_1(\mathbf{d}) + \dots - \sigma_1(\mathbf{a})\cdot\gamma_{n-1}(\mathbf{d}) + \gamma_n(\mathbf{d})\Big)$$

where

(8.11) $$\gamma_k(\mathbf{d}) \;=\; \sum (-1)^{p+|j|}\cdot s_k(\mathbf{d_j})$$

(summed over nonempty $\mathbf{j} \subseteq \{1, \dots, p\}$)

To complete the proof it is sufficient to evaluate $\gamma_k(\mathbf{d})$. We shall do so in the next lemma by giving another expression for it as a summand of $\sigma_k(\mathbf{d})$, from which its form will be immediate.

Lemma 8.12 :
$$\gamma_k(\mathbf{d}) \;=\; \begin{cases} (-1)^{p+1} & \text{if } k = 0 \\ \sigma_p(\mathbf{d})\cdot s_{k-p}(\mathbf{d}) & \text{if } k > 0 \end{cases}$$

where $s_m(\mathbf{d}) = 0$ *if* $m < 0$, *and* $= 1$ *if* $m = 0$.

Proof (of Lemma 8.12) : First if $k = 0$, then $\gamma_0(\mathbf{d})$ equals $(-1)^p\cdot\sum (-1)^{|j|}$ where the sum is over all nonempty $\mathbf{j} \subseteq \{1, \dots, p\}$. The sum is always -1 so $\gamma_0(\mathbf{d})$ equals $(-1)^{p+1}$.

Next, let $k > 0$. We give an alternate description of $\gamma_k(\mathbf{d})$. If $\mathbf{j} = \{j_1, \dots, j_r\} \subseteq \{1, \dots, p\}$, let $\gamma'_k(\mathbf{d_j})$ denote the sum of the terms in $s_k(\mathbf{d})$ which involve exactly $\{d_{j_1}, \dots, d_{j_r}\}$, i.e. only these d_{j_i} appear but each appears to at least the first power. We observe that

(8.13)

i) $$\gamma'_k(\mathbf{d_j}) \;=\; d_{j_1}\cdot d_{j_2}\cdot \dots \cdot d_{j_r}\cdot s_{k-r}(\mathbf{d_j})$$

ii) $$s_k(\mathbf{d}) \;=\; \sum \gamma'_k(\mathbf{d_j}) \quad \text{(summed over nonempty } \mathbf{j} \subseteq \{1, \dots, p\})$$

(if $r > k$ then both sides of i) are 0).

We observe that (8.11) allows us to define $\gamma_k(\mathbf{d_j})$ for any $\mathbf{j} \subseteq \{1,\dots, p\}$ by

(8.14) $$\gamma_k(\mathbf{d_j}) \;=\; \sum (-1)^{|j'|+|j|}\cdot s_k(\mathbf{d_{j'}}) \quad \text{(summed over nonempty } \mathbf{j'} \subseteq \mathbf{j}).$$

We claim that $\gamma_k(\mathbf{d_j}) = \gamma'_k(\mathbf{d_j})$ for all nonempty $\mathbf{j}$. Then, as $\gamma'_k(\mathbf{d}) = d_1 \cdot d_2 \cdot \ldots \cdot d_p \cdot s_{k-p}(\mathbf{d})$, the result follows.

The proof is by induction on $r = |\mathbf{j}|$. If $r = 1$, $\gamma'_k(d_{j_1}) = (d_{j_1})^k = \gamma_k(d_{j_1})$. Suppose it is true for each $\mathbf{j}'$ with $|\mathbf{j}'| < r$, and let $|\mathbf{j}| = r$. Then, applying ii) of (8.13) to $\mathbf{d_j}$

(8.15) $$s_k(\mathbf{d_j}) = \sum \gamma'_k(\mathbf{d_{j'}}) \quad \text{(summed over nonempty } \mathbf{j}' \subseteq \mathbf{j}\text{)}.$$

By the induction hypothesis, we may replace in (8.15) each $\gamma'_k(\mathbf{d_{j'}})$ (except for $\gamma'_k(\mathbf{d_j})$) by the expression for $\gamma_k(\mathbf{d_{j'}})$ from (8.14). We can collect together terms involving $s_k(\mathbf{d_{j'}})$ arising from $\gamma_k(\mathbf{d_{j''}})$ with $\mathbf{j}' \subseteq \mathbf{j}'' \subsetneq \mathbf{j}$. The coefficient of $s_k(\mathbf{d_{j'}})$ will be

$$\sum (-1)^{|j'|+|j''|} \quad (= (-1)^{|j'|} \cdot \sum (-1)^{|j''|}) \qquad \text{summed over } \mathbf{j}' \subseteq \mathbf{j}'' \subsetneq \mathbf{j}.$$

This equals $(-1)^{|j'|+|j|+1}$. Moving all of the terms $(-1)^{|j'|+|j|+1} s_k(\mathbf{d_{j'}})$ to the LHS of (8.15) yields

$$\gamma'_k(\mathbf{d_j}) = \sum (-1)^{|j'|+|j|} \cdot s_k(\mathbf{d_{j'}}) \quad \text{(summed over nonempty } \mathbf{j}' \subseteq \mathbf{j}\text{)}.$$

Via (8.14) we conclude $\gamma'_k(\mathbf{d_j}) = \gamma_k(\mathbf{d_j})$, completing the induction step and the proofs of the lemma and the theorem. □

Isolated Complete Intersections on Almost Free Divisors

Suppose that $V \subset \mathbb{C}^n$ is an almost free divisor and that $f: \mathbb{C}^n,0 \longrightarrow \mathbb{C}^p,0$ defines an ICIS. Under appropriate conditions, the restriction $f|V : V,0 \longrightarrow \mathbb{C}^p,0$ defines an isolated singularity on V in the sense that $f^{-1}(0)$ is (geometrically) transverse to V off 0. It has singular Milnor number $\mu(f|V) = \mu(V \cap f^{-1}(0))$.

To obtain formulas for $\mu(f|V)$, we suppose that for $f = (f_1, \ldots, f_p)$, $V_i = f_i^{-1}(0)$ is an isolated hypersurface singularity and $V, V_1, \ldots, V_p$ are in general position off 0. Then, $V' = V \cap f^{-1}(0)$ is a (geometrically) almost free complete intersection. Again for any $\mathbf{j} = \{j_1, \ldots, j_r\} \subseteq \{1, \ldots, p\}$, we can define $f_{\mathbf{j}} = (f_{j_1}, \ldots, f_{j_r})$. Then, $f_{\mathbf{j}}$ still defines an ICIS; and by lemma 7.2, $V'_{\mathbf{j}} = V \cap f_{\mathbf{j}}^{-1}(0)$ is also a (geometrically) almost free complete intersection. Thus, the $f_{\mathbf{j}}$ have the same properties as f. Hence, we refer to f but what we say also applies to the $f_{\mathbf{j}}$.
Moreover, if $V, V_1, \ldots, V_p$ are in algebraic general position off 0, then $V \cup V_{\cup\mathbf{j}}$ is

an almost free divisor.

We will compute a relative version of $\mu(f|V)$.

Define $$\delta(f|V) \stackrel{\text{def}}{=} \mu(f|V) + \mu(f).$$

This is the number of *relative vanishing cycles* on the Milnor fiber of f. More specifically, by the parametrized transversality theorem, for almost all ε and $|t| > 0$ sufficiently small, $V_t = f^{-1}(t) \cap B_\varepsilon$ is transverse to V on B_ε. Thus, $V'_t = f^{-1}(t) \cap V \cap B_\varepsilon$ is the singular Milnor fiber for V′. Since $V_t = f^{-1}(t) \cap B_\varepsilon$ is the Milnor fiber for f, it also is homotopy equivalent to a bouquet of (n−p)-spheres. Thus, from the exact sequence for a pair (V_t, V'_t), we have $H^k(V_t, V'_t) = 0$ for $k \neq n-p$ and

(8.16) $$\dim H^{n-p}(V_t, V'_t) = \mu(V') + \mu(f) = \delta(f|V).$$

This is just the analogue of the relative Milnor fiber in [Lê3] (see also [L, chap 5]). In §9 we shall give another general formula which extends that of Lê-Greuel.

For example, Orlik and Terao [OT2] has shown that for arrangements on isolated hypersurface singularities, the formulas for (8.16) take an especially simple form (see below). In our case, we obtain the following formulas for calculating $\delta(f|V)$ either in terms of the singular Milnor numbers $\mu(V \cup V_{\cup \mathbf{j}})$ or inductively from $\delta(f_{\mathbf{j}}|V)$.

Proposition 8.17 : *Suppose* $f = (f_1, \dots, f_p) : \mathbb{C}^n,0 \to \mathbb{C}^p,0$ *defines an ICIS ; and* $V_i = f_i^{-1}(0)$, $1 \le i \le p$, *together with the almost free divisor* V *are in algebraic general position off* 0. *Then,*

i) $$\delta(f|V) = \sum (-1)^{|\mathbf{j}|+p}\mu(V \cup V_{\cup \mathbf{j}})$$

where the sum is over all $\mathbf{j} = \{j_1, \dots, j_r\} \subseteq \{1, \dots, p\}$ *(even the empty set);*
or alternatively, $\delta(f|V)$ *can be computed inductively from the* $\delta(f_{\mathbf{j}}|V)$ *by*

ii) $$\delta(f|V) = \mu(V \cup (\cup V_i)) - \mu(V) - \sum \delta(f_{\mathbf{j}}|V)$$

where the sum is over all non-empty $\mathbf{j} = \{j_1, \dots, j_r\} \subsetneq \{1, \dots, p\}$.

Proof : Each part of this proposition corresponds to one of the two forms in Theorem 2. First, for i) we apply i) of Theorem 2 to $V' = V \cap V_1 \cap \dots \cap V_p$. We express the RHS in the formula for $\mu(V \cap (\cap V_i))$ as

$$\sum (-1)^{p+1+|j|+1}\mu(V \cup V_{\cup j}) + \sum (-1)^{p+1+|j|}\mu(V_{\cup j}) + (-1)^{p+1+1}\cdot\mu(V)$$

where the sum is over all non-empty $\mathbf{j} = \{j_1, \dots , j_r\} \subseteq \{1, \dots , p\}$

Again i) of Theorem 2 applied to $V_1 \cap \dots \cap V_p$ implies that the middle sum is $-\mu(\cap V_i)$. Moving this to the other side gives i).

For ii), we let $V_0 = V$ and apply ii) of Theorem 2 to $V' = V_0 \cap V_1 \cap \dots \cap V_p$. We observe that, with the exception of $\{0\}$, non-empty $\mathbf{j}' = \{j_1, \dots , j_r\} \subseteq \{0, 1, \dots , p\}$ occur in pairs $\mathbf{j}$ and $\mathbf{j} \cup \{0\}$ for non-empty $\mathbf{j} = \{j_1, \dots , j_r\} \subseteq \{1, \dots , p\}$. The sum of the terms $\mu(V_{\cap \mathbf{j}}) + \mu(V_{\cap(\mathbf{j} \cup \{0\})})$ is exactly $\delta(f_{\mathbf{j}}|V)$. Thus, the result follows by rearranging the collected terms in ii) of Theorem 2. □

For example, for p = 1 and 2 these take the forms

$$\textbf{(8.18)}\quad \delta(f|V) = \begin{cases} \mu(V \cup V_1) - \mu(V) & p = 1 \\ \mu(V \cup V_1 \cup V_2) - \mu(V) - \delta(f_1|V) - \delta(f_2|V) & p = 2 \\ \mu(V \cup V_1 \cup V_2) - \mu(V \cup V_1) - \mu(V \cup V_2) + \mu(V) & p = 2 \end{cases}$$

Arrangements on Milnor Fibers of an ICIS :

We consider the case of an almost free arrangement $A' \subset \mathbb{C}^n$ based on the free arrangement A via $\varphi : \mathbb{C}^n,0 \to \mathbb{C}^{p'},0$ and an ICIS V_0 defined by

$f_0: \mathbb{C}^n,0 \to \mathbb{C}^p,0$ with $V_0 \ \mathring{\pitchfork}_{alg}\ A'$. Thus, $V_0 \cap A'$ is an almost free complete intersection. For ε and $|t| > 0$ sufficiently small, $V_t = f_0^{-1}(t) \cap B_\varepsilon$ is transverse to A' on B_ε; and $A_t'' = f_0^{-1}(t) \cap A' \cap B_\varepsilon$, which is the singular Milnor fiber for $V_0 \cap A'$, can also be thought of as an arrangement on the Milnor fiber V_t.

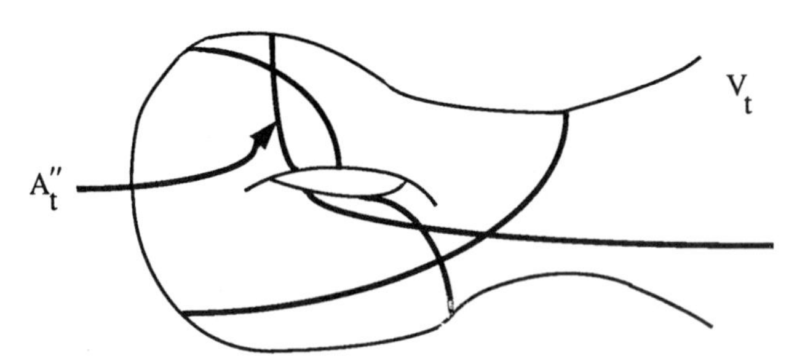

Suppose that for $f = (f_1, \dots , f_p)$, each $V_i = f_i^{-1}(0)$ is a homogeneous

isolated hypersurface singularity of degree d_i, and $A', V_1, \dots, V_p$ are in algebraic general position off 0, for an almost free arrangement A',.

Theorem 8.19 : *Suppose as above* A' *is an almost free arrangement and* f_0 *defines a homogeneous isolated complete intersection singularity* $V = \cap V_i$, *where* $A', V_1, \dots, V_p$ *are in algebraic general position off* 0, *then*

$$\textbf{(8.20)} \quad \dim H^{n-p}(V_t, A_t'') = \delta(f_0 | A') = d \cdot \Big(\sum_{j=0}^{n-p} \mu_{n-p-j}(A') \cdot s_j(\mathbf{d-1}) \Big)$$

and recall $\mu_k(A') = \sigma_k(\exp'(A))$. *Here* $d = \prod_{i=1}^{p} d_i \; (= \sigma_p(\mathbf{d}))$, $\mathbf{d-1} = (d_1-1, \dots, d_p-1)$ *(and* $\mu_0(A') = s_0(\mathbf{d-1}) = 1$).

We give the proof at the end of this section.

In the special case of a homogeneous hypersurface singularity of degree d+1, the above together with theorem 1 allows us to recover the formula obtained by Orlik and Terao [OT2].

$$\delta(f_0 | A') = (d+1) \cdot \Big(\sum_{j=0}^{n-1} \sigma_{n-1-j}(\exp'(A)) \cdot d^j \Big) = d^n \cdot P(A', d^{-1}) .$$

They have obtained this formula for arbitrary essential arrangements, while our result is restricted to almost free arrangements. However, their method depends on f_0 being a homogeneous hypersurface germ. In fact, we can obtain this formula by a more direct argument using proposition 8.17, which allows us to compute $\delta(f_0 \,|\, A')$ for arbitrary f_0.

Corollary 8.21 : *If* A' *is an almost free arrangement and* f_0 *defines an isolated hypersurface singularity* V_0 *transverse to* A' *off* 0, *then the singular Milnor number for the induced arrangement on the Milnor fiber* V_t *is given by*

$$\delta(f_0 \,|\, A') = \mu(V_0 \cup A') - \mu_n(A').$$

Since $\mu_n(A') = \sigma_n(\exp'(A))$, the only unknown term is $\mu(V_0 \cup A')$.

However, $V_0 \cup A'$ is a nonlinear arrangement based on the free arrangement

$B = A \uplus \{0\}$. Thus, if $g = (\varphi, f_0)$ then

$$\mu(V_0 \cup A') = \nu_B(g).$$

If for example, f_0 is homogeneous of degree d+1, by proposition 6.15

$$\begin{aligned}\mu(V_0 \cup A') &= \mathcal{P}_n(\exp(A))(d)\\ &= d^n + \sigma_1(\exp(A))\cdot d^{n-1} .. + \sigma_{n-1}(\exp(A))\cdot d + \sigma_n(\exp(A)).\end{aligned}$$

Since $\sigma_n(\exp(A)) - \sigma_n(\exp'(A)) = \sigma_{n-1}(\exp'(A))$, Theorem 1 again gives

$$\delta(f_0 | A') = d^n \cdot P(A', d^{-1}).$$

However, we can also apply this argument for nonhomogeneous f_0.

Example (8.22) : For example, for a Boolean arrangement A′ on the Milnor fiber of a Pham-Brieskorn polynomial given by $f_0 = x_1^{b_1} + ... + x_n^{b_n}$, $\mu_n(A') = 0$ and $\mu(V_0 \cup A')$ is computed by ii) of proposition 6.15 to be $\prod b_i$, which is then $\delta(f_0 | A')$.

The formula (8.20) makes sense for any central arrangement and we can determine the higher multiplicities from $P(A', t)$ by lemma 5.8: $P(A', t) = (1 + t)\cdot\mu(t)$. Hence, it is reasonable to conjecture

Conjecture (8.23) : For any essential arrangement A′, and a homogeneous ICIS defined by $f_0: \mathbb{C}^n,0 \to \mathbb{C}^p,0$, of degree **d**, with $V_0 = f_0^{-1}(0)$ transverse to A′ off 0, then $\delta(f_0 | A')$ is given by (8.20).

A Hypersurface on a Discriminant :

Let $g : \mathbb{C}^m,0 \to \mathbb{C}^n,0$ be a finitely determined germ with stable unfolding $G : \mathbb{C}^{m'},0 \to \mathbb{C}^{n'},0$ ($m' = m+q$ and $n' = n+q$). Let $D(g) \subset \mathbb{C}^n$ denote the discriminant of g and D(G) that of G. We consider the case $n < \hbar(D(G))$. Then, by [D2], D(g) is an almost free divisor based on D(G) via $i : \mathbb{C}^n,0 \hookrightarrow \mathbb{C}^{n'},0$. By proposition 8.17 we can inductively compute the singular Milnor fiber of $f_0 | D(g)$ for $f_0: \mathbb{C}^n,0 \to \mathbb{C}^p,0$ defining an ICIS. For the simplest case of $f_0: \mathbb{C}^n,0 \to \mathbb{C},0$ defining an isolated hypersurface singularity V_0 with $V_0 \mathring{\pitchfork}_{alg} D(g)$, again $V_0 \cap D(g)$ is an almost free complete intersection.

The computation of $\delta(f_0 | D(g))$ can again be carried out using theorem 2, specifically (8.6). Then, for $\psi = (i, f_0)$ we can repeat the argument in proposition 6.15 with $A' = D(g)$ and $A = D(G)$. Then, (6.17) becomes

(8.24) $\mu(V_0 \cup D(g)) = \nu_B(\psi) = \dim_{\mathbb{C}}(\mathcal{O}_{\mathbb{C}^n,0})^{q+1}/M$ where

$$M = \mathcal{O}_{\mathbb{C}^n,0}\{(\zeta_0', -f_0-\zeta_0''(f_0)), (\zeta_1', -\zeta_1''(f_0)), \dots, (\zeta_{n'-1}', -\zeta_{n'-1}''(f_0))\}$$

(now $\{\zeta_0, \zeta_1, \dots, \zeta_{n'-1}\}$ generate Derlog(D(G)), and ζ_0 *is* the "Euler vector field"). In the special case where $g = G$ is itself stable, $D(g) = D(G)$, so $\mu(D(g)) = 0$; and by (8.6) $\delta(f_0 | D(g)) = \mu(V_0 \cup D(g))$. This yields

Proposition 8.25 : *Let* f_0 *define an isolated hypersurface singularity* V_0 *which is transverse off* 0 *to* D(g), *the discriminant of a simple stable germ* g. *Then,*

$$\delta(f_0 | D(g)) = \dim_{\mathbb{C}}\left(\mathcal{O}_{\mathbb{C}^n,0}/(f_0+\zeta_0(f_0), \zeta_1(f_0), \dots, \zeta_{n-1}(f_0))\right)$$

where $\{\zeta_0, \zeta_1, \dots, \zeta_{n-1}\}$ *generate* Derlog(D(G)), *and* ζ_0 *is the "Euler vector field".*

Remark : The expression for $\delta(f_0 | D(g))$ is almost identical to the ring $\mathcal{O}_{\mathbb{C}^n,0}/\Theta_{X,0}\cdot f_0$ (with $X = D(G)$) appearing in [BR] (there $\zeta_0(f_0)$ is used in place of $f_0+\zeta_0(f_0)$). In certain cases such as f_0 being weighted homogeneous with the same weights as D(G), $\zeta_0(f_0) = c\cdot f_0$ and the rings are the same; however, they "should" always have the same dimension.

Proof (of Theorem 8.19) : We will apply proposition 8.17. Thus we must compute $(-1)^{|\mathbf{j}|+p}\mu(A'\cup V_{\cup\mathbf{j}})$ and sum over all $\mathbf{j} = \{j_1, \dots, j_r\} \subseteq \{1, \dots, p\}$. For empty $\mathbf{j}$ we have $(-1)^p\mu_n(A')$.

Step 1: For nonempty $\mathbf{j} = \{j_1, \dots, j_r\}$, we let $\mathbf{d_j} = (d_{j_1}, \dots, d_{j_r})$. We claim

(8.26) $\mu(A'\cup V_{\cup\mathbf{j}}) = \sigma_n(\mathbf{e}') + \sigma_{n-1}(\mathbf{e}')\cdot s_1(1^{p'-n}, \mathbf{d_j}) + \dots + s_n(1^{p'-n}, \mathbf{d_j})$

which we observe is a linear combination of $\sigma_k(\mathbf{e}')$ with coefficients depending on $\mathbf{d_j}$ where $\mathbf{e}' = (e_0-1, \dots, e_{p'-1}-1)$ for $(e_0, \dots, e_{p'-1})$ the exponents of A.

For nonempty $\mathbf{j}$, $A'\cup V_{\cup\mathbf{j}}$ is a transverse union of almost free divisors so we can apply proposition 7.12. Let $\{\zeta_0, \zeta_1, \ldots, \zeta_{p'-1}\}$ be a set of homogeneous generators for Derlog(A), with ζ_0 denoting the Euler vector field (and exp(A) = $(e_0, \ldots, e_{p'-1})$ denoting the exponents). By (7.11) and remark (7.15), a set of generators for Derlog(H) is given by

$$\text{(8.27)} \qquad \{\zeta_1, \ldots, \zeta_{p'-1}, \zeta_0 - y_1\cdot\frac{\partial}{\partial y_1}, \ldots, \zeta_0 - y_p\cdot\frac{\partial}{\partial y_p}\}.$$

Since $A' \subset \mathbb{C}^n$ is based on the free arrangement A via $\varphi : \mathbb{C}^n,0 \longrightarrow \mathbb{C}^{p'},0$, we may assume φ denotes inclusion of the first n coordinates $\varphi(x_1, \ldots, x_n) = (x_1, \ldots, x_n, 0, \ldots, 0)$). Then, projection off the first n coordinates induces an isomorphism

$$\left(\mathcal{O}_{\mathbb{C}^n,0}\right)^{p'+p} / \left(\mathcal{O}_{\mathbb{C}^n,0}\left\{(\frac{\partial}{\partial x_1}, \frac{\partial f_0}{\partial x_1}), \ldots, (\frac{\partial}{\partial x_n}, \frac{\partial f_0}{\partial x_n})\right\} \simeq \left(\mathcal{O}_{\mathbb{C}^n,0}\right)^{p'-n+p}$$

Under this isomorphism, $\zeta_1 \circ \varphi$ projects onto $\zeta'_1 \circ \varphi$, and $\zeta_0 - y_i\cdot\frac{\partial}{\partial y_i}$ onto $f_{0i}\cdot\frac{\partial}{\partial y_i}$.

Thus, by proposition 7.12,

$$\mu(A'\cup V_{\cup}) = \dim_{\mathbb{C}} \left(\mathcal{O}_{\mathbb{C}^n,0}\right)^{p'-n+p} / \left(\mathcal{O}_{\mathbb{C}^n,0}\left\{\zeta'_1 \circ \varphi, \ldots, \zeta'_{p'-1} \circ \varphi, f_{01}\cdot\frac{\partial}{\partial y_1}, \ldots, f_{0p}\cdot\frac{\partial}{\partial y_p}\right\}\right).$$

This is given by Theorem 2 of [D4] as the Macaulay-Bezout number $B(\mathbf{b}: 1^n, \mathbf{c})$ for the weights $\mathbf{b} = (\mathbf{e}', 0^p)$ and $\mathbf{c} = (1^{p'-n}, \mathbf{d})$ where again $\mathbf{e}' = (e_0-1, \ldots, e_{p'-1}-1)$. We obtain

$$\begin{aligned}\mu(A'\cup V_{\cup}) &= \sigma_n(\mathbf{e}', 0^p) + \sigma_{n-1}(\mathbf{e}', 0^p)\cdot s_1(1^{p'-n}, \mathbf{d}) + \ldots + s_n(1^{p'-n}, \mathbf{d}) \\ &= \sigma_n(\mathbf{e}') + \sigma_{n-1}(\mathbf{e}')\cdot s_1(1^{p'-n}, \mathbf{d}) + \ldots + s_n(1^{p'-n}, \mathbf{d}).\end{aligned}$$

This can be repeated for $\mu(A'\cup V_{\cup \mathbf{j}})$ by replacing f_0 by $f_{0\mathbf{j}}$. We obtain (8.26) establishing the first step.

<u>Step 2</u> : By proposition 8.17, we multiply (8.26) by $(-1)^{|\mathbf{j}|+p}$ and sum over all $\mathbf{j} \subseteq \{1, \ldots, p\}$. Hence, when we sum over $\mathbf{j}$, we obtain a linear combination

$$\text{(8.28)} \quad \delta(f|V) = \beta_0(\mathbf{d})\sigma_n(\mathbf{e}') + \beta_1(\mathbf{d})\cdot\sigma_{n-1}(\mathbf{e}') + \ldots + \beta_n(\mathbf{d})\cdot\sigma_0(\mathbf{e}') + (-1)^p\mu_n(A')$$

where $(-1)^p\mu_n(A')$ has been separated so we only have to consider nonempty $\mathbf{j}$, and

$$\text{(8.29)} \qquad \beta_k(\mathbf{d}) = \sum (-1)^{|\mathbf{j}|+p} s_k(1^{p'-n}, \mathbf{d_j})$$

summed over <u>nonempty</u> $\mathbf{j} \subseteq \{1, \ldots, p\}$.

Step 3 : We claim

$$\beta_k(\mathbf{d}) = (-1)^{p+1}\cdot s_k(1^{p'-n}) + \sigma_p(\mathbf{d})\cdot\Big(\sum_{\ell=1}^{k-p} s_{k-p-\ell}(1^{p'-n})\cdot s_\ell(\mathbf{d})\Big) \tag{8.30}$$

To evaluate (8.29), we may expand $s_k(1^{p'-n}, \mathbf{d}_j)$ by the product rule

$$s_k(1^{p'-n}, \mathbf{d}_j) = \sum_{\ell=0}^{k} s_{k-\ell}(1^{p'-n})s_\ell(\mathbf{d}_j) \tag{8.31}$$

Thus, using (8.31) for (8.29) we obtain

$$\beta_k(\mathbf{d}) = \sum_{\ell=0}^{k} s_{k-\ell}(1^{p'-n})\cdot\Big(\sum(-1)^{|j|+p}s_\ell(\mathbf{d}_j)\Big)$$

$$= \sum_{\ell=0}^{k} s_{k-\ell}(1^{p'-n})\cdot\gamma_\ell(\mathbf{d}) \tag{8.32}$$

with $\gamma_\ell(\mathbf{d})$ defined as in lemma 8. 12. By that lemma

$$\beta_k(\mathbf{d}) = (-1)^{p+1}\cdot s_k(1^{p'-n}) + \sigma_p(\mathbf{d})\cdot\Big(\sum_{\ell=1}^{k} s_{k-\ell}(1^{p'-n})\cdot s_{\ell-p}(\mathbf{d})\Big) \tag{8.33}$$

which yields (8.30) on replacing ℓ by $\ell-p$ (since $s_m(\mathbf{d}) = 0$ if $m < 0$).

Substituting into (8.28)

$$\delta(f|V) = (-1)^p\mu_n(A') + \sum_{k=0}^{n}\beta_k(\mathbf{d})\cdot\sigma_{n-k}(\mathbf{e}')$$

$$= (-1)^p\mu_n(A') + (-1)^{p+1}\cdot\Big(\sum_{k=0}^{n} s_k(1^{p'-n})\cdot\sigma_{n-k}(\mathbf{e}')\Big) + \tag{8.34}$$

$$\sigma_p(\mathbf{d})\cdot\Big(\sum_{k=0}^{n}\sum_{\ell=0}^{k-p}\Big(s_{k-p-\ell}(1^{p'-n})\cdot s_\ell(\mathbf{d})\Big)\cdot\sigma_{n-k}(\mathbf{e}')\Big)$$

Step 4 : We claim the sum of the first and middle sum of (8.34) is zero.

The sum in the second term can be evaluated using proposition 2.19 of [D4]. In the notation of (2.18) of [D4]

$$\tau(D(\mathbf{e}'', 1^{p'-n})) = \sum_{k=0}^{n} s_k(1^{p'-n})\cdot\sigma_{n-k}(\mathbf{e}') ;$$

where $\mathbf{e}'' = (e_1-1, \dots , e_{p'-1}-1)$ (and $\sigma_r(\mathbf{e}'') = \sigma_r(0, \mathbf{e}') = \sigma_r(\mathbf{e}')$). The matrices

$$D(\mathbf{e}'', 1^{p'-n}) = D(\mathbf{e}, 0^{p'-n})$$

where again $\mathbf{e} = (e_1, \dots , e_{p'-1}) = \exp'(A)$. Again applying proposition 2.19 of [D4] to $D(\mathbf{e}, 0^{p'-n})$

$$\tau(D(\mathbf{e}, 0^{p'-n})) = \sigma_n(\mathbf{e}) \quad (= \sigma_n(\exp'(A))).$$

By Theorem 1, $\mu_n(A') = \sigma_n(\exp'(A))$; hence the first and second terms of (8.32) cancel.

Step 5 : Lastly, to complete the proof, we claim the last term equals the desired formula. By the product formula for s_k,

$$s_{k-p}(1^{p'-n}, \mathbf{d}) = \sum_{\ell=0}^{k-p} s_{k-p-\ell}(1^{p'-n})\cdot s_\ell(\mathbf{d}) .$$

Thus, the third term becomes

$$\sigma_p(\mathbf{d})\cdot\Big(\sum_{k=0}^{n} s_{k-p}(1^{p'-n}, \mathbf{d})\cdot\sigma_{n-k}(\mathbf{e}') \Big) . \tag{8.35}$$

Again since $s_m(1^{p'-n}, \mathbf{d}) = 0$ if $m < 0$, we can replace k by k−p so (8.35) becomes

$$\sigma_p(\mathbf{d})\cdot\Big(\sum_{k=0}^{n-p} s_k(1^{p'-n}, \mathbf{d})\cdot\sigma_{n-p-k}(\mathbf{e}') \Big) . \tag{8.36}$$

To rewrite this, we apply proposition 2.19 of [D4] one last time.

$$\tau(D(\mathbf{e}'', (1^{p'-n}, \mathbf{d}))) = \sum_{k=0}^{n-p} s_k(1^{p'-n}, \mathbf{d})\cdot\sigma_{n-p-k}(\mathbf{e}') ;$$

but
$$D(\mathbf{e}'', (1^{p'-n}, \mathbf{d})) = D(\mathbf{e}, (0^{p'-n}, \mathbf{d}-1))$$

and
$$\tau(D(\mathbf{e}, (0^{p'-n}, \mathbf{d}-1))) = \sigma_p(\mathbf{d})\cdot\Big(\sum_{j=0}^{n-p} \sigma_{n-p-j}(\exp'(A))\cdot s_j(\mathbf{d}-1) \Big).$$

Since $\mu_k(A') = \sigma_k(\exp'(A))$ by proposition 5.2, this completes the proof of the theorem. □

§9 Relative Milnor Fibers and the Lê-Greuel Formula

We gave in §8 a number of formulas for the singular Milnor numbers and higher multiplicities for almost free complete intersections by representing them as transverse intersections of almost free divisors. This allowed us to use the Bezout theorem for determinantal modules to give specific formulas in the weighted homogeneous case. In the case of an ICIS, there is a very different formula due to Lê-Greuel [Lê3] and [Gr], for the Milnor number as an alternating sum of relative Milnor numbers, where the relative Milnor number is given explicitly as the length of an algebra (see also Teissier [Te,part II] for a simpler case). These two approaches for computing the Milnor number are possible because the smooth and "singular Milnor fibers" for an ICIS are the same.

However, for nonisolated complete intersections, these two notions differ. Now, the correct version of this formula applies to the singular Milnor fiber of an almost free divisor on the smooth Milnor fiber of the nonisolated complete intersection. It expresses the relative Euler characteristic as the length of a determinantal module. In the case of an almost free divisor on an ICIS, this does give an alternate method to compute the relative singular Milnor number, and recovers a module version of the Lê-Greuel formula.

Consider almost free divisors $V_i \subset \mathbb{C}^n$, $1 \leq i \leq k$, which are based on free divisors $V_i' \subset \mathbb{C}^{p_i}$ via $\varphi_i : \mathbb{C}^n,0 \to \mathbb{C}^{p_i},0$. Let $p = \sum p_i$. and $p' = p - p_k$. Suppose

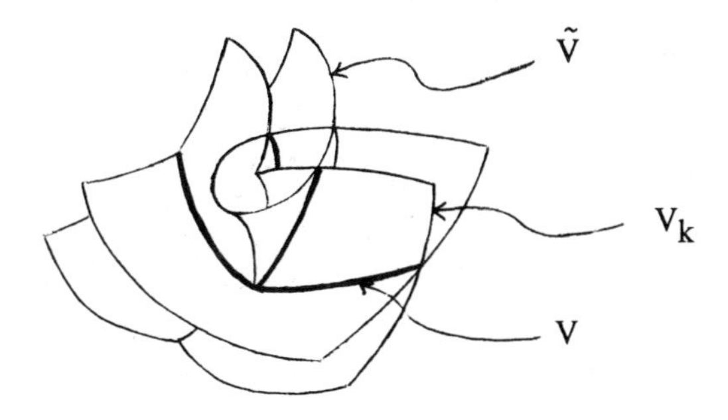

the V_i are in algebraic general position off 0, and define $V = \cap V_i$ and $\tilde{V} = \bigcap_{i=1}^{k-1} V_i$. Both V and $\tilde{V}$ are almost free complete intersections. *Throughout this section we assume* $n < \min\{\tilde{h}(V_i)\}$; this ensures that algebraic and geometric transversality coincide for the intersection of $\mathbb{C}^n$ with any collection of the V_i.

If we let $V' = \prod V_i'$ and $\tilde{V}' = \prod_{i=1}^{k-1} V_i'$, then $V = \varphi^{-1}(V')$ and $\tilde{V} = \varphi'^{-1}(\tilde{V}')$ for $\varphi = (\varphi_1, \ldots, \varphi_k)$ and $\varphi' = (\varphi_1, \ldots, \varphi_{k-1})$. Let $h_i : \mathbb{C}^{p_i},0 \longrightarrow \mathbb{C},0$ be a good defining equation for V_i'. For sufficiently general c_i, let $V_{is}' = h_i^{-1}(c_i \cdot s)$; then $V_{is}' \cap B_\varepsilon$ is the smooth Milnor fiber of V_i' for sufficiently small s and $\varepsilon > 0$.

We consider a stabilization $\varphi_t : U \longrightarrow \mathbb{C}^p,0$ of $\varphi = (\varphi_1, \ldots, \varphi_k)$ relative to the product union $\uplus V_i'$. so that φ_0 is a representative of φ, algebraically transverse off 0. We require that $\varphi'_t = (\varphi_{1t}, \ldots, \varphi_{k-1,t}) : U \longrightarrow \mathbb{C}^{p'},0$ is also a stabilization of $\varphi' = (\varphi_1, \ldots, \varphi_{k-1})$ and φ_{kt} is a stabilization of φ_k for small $t \neq 0$. Finally we may suppose that for $|t| << s$, φ'_t is transverse to $\tilde{V}_s' = \prod_{i=1}^{k-1} V_{is}'$ and φ_t is transverse to $V_s' = \tilde{V}_s' \times V_k'$.

Then, given $\varepsilon > 0$ such that $B_\varepsilon \subset U$, we may choose $|s| < \gamma$ sufficiently small so that the intersections of B_ε with $\tilde{V}_{t,s} = \varphi'_t{}^{-1}(\tilde{V}_s')$ and $V_{kt} = \varphi_t^{-1}(V_k')$ are the smooth (resp.singular) Milnor fibers for $\tilde{V}$ and V_k. From now on, we assume we restrict to a sufficiently small B_ε and choose s and t sufficiently small without specifically saying so.

We are interested in a formula for the relative Euler characteristic

$$\tilde{\chi}(\tilde{V}, V) \stackrel{\text{def}}{=} \chi(\tilde{V}_{t,s} \cap B_\varepsilon, \tilde{V}_{t,s} \cap V_{kt} \cap B_\varepsilon).$$

We note that in the special case that $\tilde{V}$ is an ICIS, $\tilde{V}_{t,s}$ is the smooth Milnor fiber of $\tilde{V}$ and $V_{t,s} = \tilde{V}_{t,s} \cap V_{kt}$ is the singular Milnor fiber of V. Then, $V_{t,s}$ and $\tilde{V}_{t,s}$ are homotopy equivalent to bouquets of spheres of dimensions n−k and n−k+1; it follows that $H^j(\tilde{V}, V) = 0$ for $j \neq n-k+1$ and then, in the special case that $\tilde{V}$ is an ICIS,

$$(9.1) \qquad |\tilde{\chi}(\tilde{V}, V)| = \dim_{\mathbb{C}} H^{n-k+1}(\tilde{V}, V) = \mu(\tilde{V}) + \mu(V)$$

The Lê-Greuel formula, gives in the case when both $\tilde{V}$ and V are ICIS a formula for (9.1) as the length of an algebra. We shall give a formula applicable to almost free complete intersections which will instead be in terms of the length of a determinantal module.

For each V_i', h_i is a good defining equation. Let $\{\zeta_0^{(i)}, \zeta_1^{(i)}, \ldots, \zeta_{p_i-1}^{(i)}\}$ be a set of generators for Derlog(V_i') with $\zeta_0^{(i)}$ denoting the "Euler vector field" and the remainder generating Derlog(h_i). Also, $f_i = h_i \circ \varphi$ is the defining equation for V_i, and $I(\tilde{V})$, generated by $\{f_1, \ldots, f_{k-1}\}$, is the ideal defining the complete intersection $\tilde{V}$. Let $D(\tilde{V}, V)$ denote the quotient module of $\left(\mathcal{O}_{\mathbb{C}^n,0}\right)^p$ given by

$$(9.2) \quad \left(\mathcal{O}_{\mathbb{C}^n,0}\right)^p \Big/ \left(\mathcal{O}_{\mathbb{C}^n,0}\left\{\frac{\partial\varphi}{\partial x_i}\right\} + \mathcal{O}_{\mathbb{C}^n,0}\left\{\zeta_i^{(j)}\circ\varphi,\ i \geq 1 \text{ and } 1 \leq j \leq k\right\} + I(\tilde{V})\cdot\left(\mathcal{O}_{\mathbb{C}^n,0}\right)^p\right).$$

We observe that

$$(9.3) \qquad D(\tilde{V}, V) \simeq \left(\mathcal{O}_{\tilde{V},0}\right)^p \Big/ \left(\mathcal{O}_{\tilde{V},0}\left\{\frac{\partial\varphi}{\partial x_i}\right\} + \mathcal{O}_{\tilde{V},0}\left\{\zeta_i^{(j)}\circ\varphi,\ i \geq 1 \text{ and } 1 \leq j \leq k\right\}\right)$$

is a quotient module of $\left(\mathcal{O}_{\tilde{V},0}\right)^p$ by a submodule on $n+p-k = (n-k+1)+p-1$ generators and $\tilde{V}$ is a complete intersection (and hence Cohen-Macaulay) of dimension $n-k+1$. Thus, by classical results of Macaulay-Northcott [Mc], [No], if $\dim_{\mathbb{C}} D(\tilde{V}, V) < \infty$, then $D(\tilde{V}, V)$ is a determinantal Cohen-Macaulay module. Then, the relative Euler characteristic is given by the following formula.

Theorem 9.4 : *Suppose as* $\tilde{V}$ *and* V *are defined as above where* $V_1, \ldots, V_k$ *are almost free divisors in algebraic general position off* 0, (*and* $n < \min\{\tilde{h}(V_i)\}$). *Then*

$$\tilde{\chi}(\tilde{V}, V) = (-1)^{n-k+1} \dim_{\mathbb{C}} D(\tilde{V}, V)$$

Remark : In the case of an ICIS V defined by $\varphi : \mathbb{C}^n,0 \longrightarrow \mathbb{C}^p,0$ (here $k = p$), there will be no $\zeta_i^{(j)}$ with $i \geq 1$, so by (9.1) the formula takes the form

$$(9.5) \qquad \mu(\tilde{V}) + \mu(V) = \dim_{\mathbb{C}} \left(\mathcal{O}_{\tilde{V},0}\right)^p / \mathcal{O}_{\tilde{V},0}\left\{\frac{\partial\varphi}{\partial x_i}\right\}$$

This is the module version of the Lê-Greuel formula

$$\mu(\tilde{V}) + \mu(V) = \dim_{\mathbb{C}} \mathcal{O}_{\tilde{V},0}/J(\varphi)$$

where $J(\varphi)$ is the Jacobian ideal generated by the $p \times p$ minors of the Jacobian matrix $D\varphi$.

As a corollary, we obtain another formula for the relative homology for an arrangement A_t'' induced on the Milnor fiber V_t of the germ $f_0 : \mathbb{C}^n,0 \to \mathbb{C}^p,0$ defining an ICIS V_0 by an almost free arrangement $A' \subset \mathbb{C}^n$. Here we use the notation of §8, so A' is based on the free arrangement A via $\varphi : \mathbb{C}^n,0 \to \mathbb{C}^{p'},0$

with $V_0 \mathring{\pitchfork}_{alg} A'$. Also, $\{\zeta_0, \zeta_1, \dots, \zeta_{p'-1}\}$ generate Derlog(A), with ζ_0 the "Euler vector field" and with the remainder generating Derlog(H). Let $f = (\varphi, f_0) : \mathbb{C}^n,0 \to \mathbb{C}^{p+p'},0$. A direct application of theorem 9.4 yields

Corollary 9.6 : *Suppose as above* A' *is an almost free arrangement and* f_0 *defines an isolated complete intersection singularity* $V = \cap V_i$, *where* $A', V_1, \dots, V_p$ *are in algebraic general position off* 0, *then*

$$\dim_{\mathbb{C}} H^{n-p}(V_t, A_t'') = \dim_{\mathbb{C}} \left(\mathcal{O}_{V_0,0}\right)^{p+p'} / \left(\mathcal{O}_{V_0,0}\left\{\frac{\partial f}{\partial x_i}, \zeta_i \circ \varphi, i \geq 1\right\}\right)$$

Remark : This result applies equally well to any almost free divisor A'. Also, in the case f_0 defines an isolated hypersurface singularity and A' itself is free, then φ = id and we obtain

$$\left(\mathcal{O}_{V_0,0}\right)^{n+1} / \left(\mathcal{O}_{V_0,0}\left\{\frac{\partial f}{\partial x_i}, (\zeta_i, 0), i \geq 1\right\}\right) \simeq \left(\mathcal{O}_{\mathbb{C}^n,0}/(f_0, \zeta_1(f_0), \dots, \zeta_{n-1}(f_0))\right)$$

Hence

$$(9.7) \qquad \dim_{\mathbb{C}} H^{n-p}(V_t, A_t'') = \dim_{\mathbb{C}} \left(\mathcal{O}_{\mathbb{C}^n,0}/(f_0, \zeta_1(f_0), \dots, \zeta_{n-1}(f_0))\right)$$

which gives yet another variant of the formula from proposition 8.25.

Before beginning the proof, we outline the main ideas.

The line of the proof is actually very similar to that in [DM] except that we will be working on singular spaces. First, in [DM], we used a result of Siersma which, in turn, was an extension of a standard type Morse theory argument, see e.g. [L], which allowed us to compute the topology of a highly singular germ of a variety in terms of near-by isolated singular points. We shall further extend this lemma to a germ on a nonisolated singular space. For this we will make considerable use of algebraic transversality to keep track of singular behavior.

Second, for the stabilization of $\tilde{V}$, we shall extend the definition of $D(\tilde{V}, V)$ to a sheaf $\mathcal{M}$ on the stabilization. Then, as in [DM], using Cohen-Macaulay properties of $\mathcal{M}$, we shall show that it projects to a free sheaf on $\mathbb{C}$ giving the correct deformation properties.

Third, we show that at a point of support of $\mathcal{M}$, its length will be the relative Euler characteristic of the pair of smooth and singular Milnor fibers. This is done by applying the modified version of Siersma's result. It reduces to the case of the Milnor number of an isolated singularity on a smooth space. Here we are effectively giving a direct proof of the module version (9.5) of the Lê-Greuel formula.

Example 9.8 : It is in the third step that we must assume that the ambient Milnor fiber is smooth. In fact, this result would be false for the singular Milnor fiber. To see this, consider the product union $V = V_1 \uplus V_2$ of two cusp singularities V_1 defined by $h_1(x, y) = y^3 + x^2 = 0$, and V_2 defined by $h_2(z, w) = w^3 + z^2 = 0$. We consider a generic linear section W in $\mathbb{C}^4$ defined by $\ell(x, y, z, w) = ax + by + cz + dw = 0$. In this case, $V_1 \times V_2$ is the transverse intersection of the free divisors $V_1 \times \mathbb{C}^2$ and $\mathbb{C}^2 \times V_2$. If W projects onto each factor of $\mathbb{C}^2$, then we can compute both the relative singular Milnor number of $\ell \,|\, V_1 \times V_2$ and $\tilde{\chi}(\tilde{V}, V)$.

First, $\mathrm{Derlog}(h_1)$ is generated by $\zeta_1^{(1)} = 3y^2\frac{\partial}{\partial x} - 2x\frac{\partial}{\partial y}$ and similarly for $\mathrm{Derlog}(h_2)$, by $\zeta_1^{(2)}$ where x and y are replaced by z and w. If $i : W \hookrightarrow \mathbb{C}^4$ denotes the inclusion of W, then the singular Milnor number equals $\mu_{V_1 \times V_2}(i)$, which equals $\mu(V'_1 \cap V'_2)$ where $V'_1 = i^{-1}(V_1 \times \mathbb{C}^2)$ and similarly for V'_2. As

W projects onto each $\mathbb{C}^2$, $\mu(V'_i) = 0$ and so by theorem 2 (specifically (8.7))

$$\mu_{V_1 \times V_2}(i) = \mu(V'_1 \cap V'_2) = \mu(V'_1 \cup V'_2) = \mu_{V_1 \otimes V_2}(i)$$

By a calculation using proposition 7.12, $\mu_{V_1 \times V_2}(i) = 1$ (see fig. 9.9).

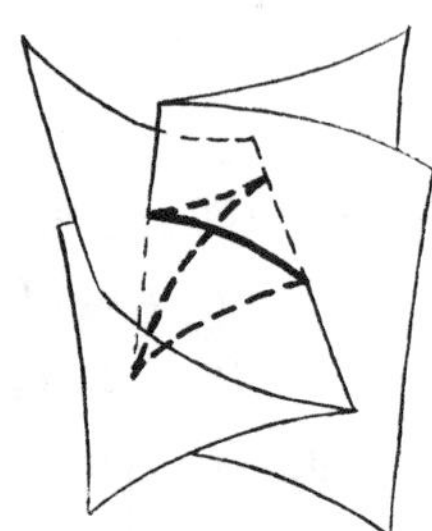

Fig. 9.9

However, by contrast, a calculation shows $|\tilde{\chi}(\tilde{V}, V)| = \dim_{\mathbb{C}} D(\tilde{V}, V) = 9$, which is the number of relative singular vanishing cycles of ℓ on the (smooth) Milnor fiber of (h_1, h_2).

Proof (of the theorem) :

Step 1 : First, we replace the parameters s and t in the stabilizations by a single parameter. For this, it is sufficient to show that

$$W = \{(s, t) \in \mathbb{C}^2 : \varphi'_t \text{ is transverse to } \tilde{V}_s' \text{ and } \varphi_t \text{, to } V_s'\}$$

is a nonempty Zariski open set (in a sufficiently small neighborhood U of 0). For then, we may choose a curve $\gamma(t') : \mathbb{C}, 0 \longrightarrow \mathbb{C}^2, 0$ with $\gamma(t') \in W$ for $t' \neq 0$. We replace (s, t) by $\gamma(t')$ so that now $V_{it'}' = h_i^{-1}(c_i \cdot \gamma_1(t'))$, $\varphi_{t'} = \varphi_{\gamma_2(t')}$, and $\varphi'_{t'} = \varphi'_{\gamma_2(t')}$ where $\gamma(t') = (\gamma_1(t'), \gamma_2(t'))$. Lastly, we replace t' by t. Then,

$$\mathcal{V}' = \{(y, t) \in \mathbb{C}^{p+1} : h_i(y) = c_i \cdot \gamma_1(t), 1 \leq i < k, \ h_k(y) = 0\} \quad \text{and}$$

$$\tilde{\mathcal{V}}' = \{(y, t) \in \mathbb{C}^{p'+1} : h_i(y) = c_i \cdot \gamma_1(t), 1 \leq i < k\}$$

are complete intersections; and the restrictions to them of the projection $(y, t) \mapsto t$ to $\mathbb{C}$ have fibers V_t', resp. $\tilde{V}_t'$.

To see that W is Zariski open, we consider

$$W' = \{(x, s, t) \in \mathbb{C}^{n+2} : \varphi'_t \text{ is transverse to } \tilde{V}_s' \text{ and } \varphi_t \text{, to } V_s' \text{ at } x\}.$$

As $n < \min\{\tilde{h}(V_i)\}$, we may use $T_{\log}H_i$ to determine transversality to V_{it}'. even when t = 0. Then, the set of points where algebraic transversality fails, i.e. S =

$\mathbb{C}^{n+2} \setminus W'$, is a Zariski closed subset and $S \cap (\mathbb{C}^n \times \{0\}) = \{0\}$. Thus, the restriction to S of projection pr_2 to $\mathbb{C}^2$ is finite to one. Hence, the image $pr_2(S)$ is also Zariski closed and its complement is W.

Step 2 : For the stabilization $\varphi_t : U \longrightarrow \mathbb{C}^p,0$, we form $\Phi : U \times \mathbb{C} \longrightarrow \mathbb{C}^p \times \mathbb{C}$ by $\Phi(x, t) = (\varphi_t(x), t)$ and define $(\mathcal{V}, 0) \subset \mathbb{C}^n \times \mathbb{C}, 0$ to be $\Phi^{-1}(\mathcal{V}')$. Then $\mathcal{V}$ is a complete intersection of dimension n-k+1. Also, we let $\pi : \mathcal{V}, 0 \longrightarrow \mathbb{C}, 0$ denote the restriction to $\mathcal{V}$ of the projection $\mathbb{C}^n \times \mathbb{C} \longrightarrow \mathbb{C}$. Then, $\pi^{-1}(t) = V_t = \varphi_t^{-1}(V_t')$ is a complete intersection at all points $x \in U$. We have a similar construction of $\tilde{\mathcal{V}}$ and $\pi : \tilde{\mathcal{V}}, 0 \longrightarrow \mathbb{C}, 0$ from φ'_t and $\tilde{\mathcal{V}}'$, with $\tilde{V}_t$ as a fiber. In this later case, we have $h' : \tilde{\mathcal{V}}, 0 \longrightarrow \mathbb{C}, 0$ which is the restriction of $h = H_k \circ \varphi_{kt}$. Then, $h^{-1}(0) \cap \tilde{V} = V$. Also, by the construction of the stabilization, $h^{-1}(0) \cap \tilde{V}_t$ is a singular Milnor fiber for V_k on the smooth Milnor fiber of $\tilde{V}$.

We define an analogue of (9.3) but for the deformation Φ. We let

$$\textbf{(9.10)} \qquad \mathcal{M} = \left(\mathcal{O}_{\tilde{\mathcal{V}},0}\right)^p / \left(\mathcal{O}_{\tilde{\mathcal{V}},0}\left\{\frac{\partial \Phi}{\partial x_i}\right\} + \mathcal{O}_{\tilde{\mathcal{V}},0}\left\{\zeta_i^{(j)} \circ \Phi,\ i \geq 1 \text{ and } 1 \leq j \leq k\right\}\right)$$

define a sheaf of modules on the complete intersection $\tilde{\mathcal{V}}$ (and hence Cohen-Macaulay) of dimension n-k+2. It is a quotient module of $\left(\mathcal{O}_{\tilde{\mathcal{V}},0}\right)^p$ by a submodule on $n+p-k = (n-k+2)+p-2$ generators. Thus, by results of Macaulay-Northcott [Mc], [No], $\dim \operatorname{supp}(\mathcal{M}) \geq 1$. However, $\mathcal{M}/t\mathcal{M} \simeq D(\tilde{V}, V)$. So if $\dim_{\mathbb{C}} D(\tilde{V}, V) < \infty$, then on $\mathbb{C}^n \times \{0\}$, this module has support at the origin. Thus, $\operatorname{supp}(\mathcal{M})$ is 1-dimensional (and hence 0-dimensional in any $\mathbb{C}^n \times \{t\}$) and Cohen-Macaulay. Thus, as in prop. 5.2 of [DM], the push-forward $\pi_*\mathcal{M}$ for the projection π: $\mathbb{C}^{n+1},0 \longrightarrow \mathbb{C},0$ is Cohen-Macaulay of dimension = 1, and hence free. Thus, $\dim_{\mathbb{C}} (\pi_*\mathcal{M} / m_{t-t_0}\pi_*\mathcal{M})$ is constant. Also, ,

$$\textbf{(9.11)} \qquad \dim_{\mathbb{C}} (\pi_*\mathcal{M} / m_{t-t_0}\pi_*\mathcal{M}) = \sum \dim_{\mathbb{C}} D(\tilde{V}_{t_0}, V)_{x_i}$$

where $D(\tilde{V}_{t_0}, V)_{x_i}$ is defined as in (9.3) except the germs are taken at x_i and Φ is replaced by φ_{t_0}. Also, this sum is over the finite set of $\operatorname{supp}(\mathcal{M})$ over t_0. Hence, summarizing, we obtain

Lemma 9.12 : *In the above situation, for sufficiently small* t_0 ,

$$\dim_{\mathbb{C}} D(\tilde{V}, V) = \sum \dim_{\mathbb{C}} D(\tilde{V}_{t_0}, V)_{x_i}$$

where the sum is over the finite set of $x_i \in \mathrm{supp}(\mathcal{M})$ *over* t_0.

<u>Step 3</u> : Our next goal is to show that the points x_i where $\dim_{\mathbb{C}} D(\tilde{V}_{t_0}, V)_{x_i} > 0$ are the points where the algebraic transversality fails for $h^{-1}(y_i)$ and $\tilde{V}_{t_0}$ (where $y_i = h(x_i)$). Moreover, we shall interpret these points as exactly the points where $h' (= h|\tilde{V}_{t_0}) : \tilde{V}_{t_0}, x_i \longrightarrow \mathbb{C}$ has an isolated singularity. Ultimately, we will interpret the dimensions $\dim_{\mathbb{C}} D(\tilde{V}_{t_0}, V)_{x_i}$ as Milnor numbers for $h' : \tilde{V}_{t_0}, x_i \longrightarrow \mathbb{C}$.

If $\dim_{\mathbb{C}} D(\tilde{V}_{t_0}, V)_x > 0$, then the vectors

$$\{\frac{\partial \varphi_{t_0}}{\partial x_1}(x), \dots, \frac{\partial \varphi_{t_0}}{\partial x_n}(x)\} \cup \left\{ \zeta_i^{(j)} \circ \varphi_{t_0}(x), i \geq 1 \text{ and } 1 \leq j \leq k \right\} \tag{9.13}$$

do not span $T\mathbb{C}^p$. The condition $n < \min\{\tilde{h}(V_i)\}$ implies for $i < k$ that $T_{\log}(H_i)_{\varphi_{it_0}(x)} = S^{(i)}{}_{j\varphi_{it_0}(x)}$, the stratum of the canonical Whitney stratification of V_i'. containing $\varphi_{it_0}(x)$. Thus, the vectors $\left\{ \zeta_i^{(j)}, i \geq 1 \text{ and } 1 \leq j < k \right\}$ span S'_i a stratum for the Whitney stratification of $\tilde{V}'$. Also, $\left\{ \zeta_i^{(k)}, i \geq 1 \right\}$ span $T_x H_k^{-1}(y)$ when $y \neq 0$, by the properties of Derlog(H). Thus, the vectors in (9.13) fail to span $T\mathbb{C}^p$ for an $x \in h^{-1}(y) \cap \tilde{V}_{t_0}$ exactly when φ_{t_0} fails to be geometrically (equivalently algebraically) transverse to $\tilde{V}_{t_0}' \times H_k^{-1}(y)$. Because φ_{t_0} , along with φ'_{t_0} and φ_{kt_0} , are stabilizations, φ_{kt_0} is transverse to V'_k and φ'_{t_0} is transverse to $\tilde{V}_{t_0}'$; hence, by the openness of transversality (to closed Whitney stratified spaces), φ_{kt_0} is transverse to $H_k^{-1}(y)$ for y sufficiently close to 0. Thus, by i) of Lemma 7.2, $h^{-1}(y) = \varphi_{kt_0} \circ H_k^{-1}(y)$ is transverse to $\tilde{V}_{t_0} = \varphi'_{t_0}{}^{-1}(\tilde{V}_{t_0}')$ iff φ_{t_0} is transverse to $\tilde{V}_{t_0}' \times H_k^{-1}(y)$.

Putting these arguments together, we see that for y sufficiently close to 0 if $x \in h^{-1}(y) \cap \tilde{V}_{t_0}$ then $h^{-1}(y) \ \overline{\pitchfork}_{alg} \ \tilde{V}_{t_0}$ at x iff $\dim_{\mathbb{C}} D(\tilde{V}_{t_0}, V)_x = 0$. Thus, algebraic transversality of $h^{-1}(y)$ to $\tilde{V}_{t_0}$ only fails at isolated points; and there are only a finite number of such points. Since algebraic transversality implies geometric transversality, it follows that the point x above is an isolated singular point for $h| \tilde{V}_{t_0}$ (recall $\tilde{V}_{t_0}$ is the smooth Milnor fiber of $\tilde{V}$).

Step 4 : Next we apply a generalization of the lemma of Siersma used in [DM]. Siersma shows that the standard Morse theory type argument, see e.g [L, chap. 5] can also be applied to non-isolated singularities defined by germs $g : \mathbb{C}^{n+1}, 0 \to \mathbb{C}, 0$. We show this extends to germs defined on (nonisolated) complete intersections.

Let $X, 0 \subset \mathbb{C}^{n+1}, 0$ be a k- dimensional complete intersection and let g, h : $\mathbb{C}^{n+1}, 0 \to \mathbb{C}, 0$ define hypersurfaces singularities $Y = g^{-1}(0)$ and $Z = h^{-1}(0)$ so that $Y \cap X$ is a complete intersection with (smooth) Milnor fiber $Y_{t,\varepsilon} = Y_t \cap X \cap B_\varepsilon$. We allow $Y \cap X \cap Z$ to have a nonisolated singularity at 0. However, we suppose that for sufficiently small γ, if $\gamma > |t|, |s| \geq 0$, then $Y_t \cap X \cap Z_s$ is (stratified) transverse to the Milnor sphere $S_\varepsilon = \partial B_\varepsilon$. Moreover, we suppose that on $Y_{t_0,\varepsilon} \setminus Z$, h has only a finite number of isolated singularities which approach 0 as t_0 approaches 0.

By step 2,this is exactly the situation we are in with $X = \tilde{\mathcal{V}}$, $g = pr : \mathbb{C}^{n+1}, 0 \to \mathbb{C}, 0$, and h as defined above. We let $Z_{s,t_0,\varepsilon} = X \cap Y_{t_0} \cap Z_s \cap B_\varepsilon$.

Proposition 9.14 : *In the above situation, with* $|s| < \delta$, $Y_{t_0,\varepsilon}$ *is homotopy equivalent to the spaces obtained from* $Z_{0,t_0,\varepsilon}$ $(= X \cap Y_{t_0} \cap Z \cap B_\varepsilon)$ *by adjoining cells of dimension* k-1, *and the number of such cells equals the sum of the Milnor numbers of the isolated singularities of* h *on* $(X \cap Y_{t_0}) \setminus Z$.

Hence, $\chi(Y_{t_0,\varepsilon}, Z_{0,t_0,\varepsilon})$ = *sum of the Milnor numbers of* $h| (X \cap Y_{t_0}) \setminus Z$

We first use this result to complete the proof of the theorem.

Step 5 : To prove the theorem, by steps 3 and 4, it is sufficient to prove that at each isolated singular point x of $h | (\tilde{V}_{t_0} \setminus Z)$, $\dim_{\mathbb{C}} D(\tilde{V}_{t_0}, V)_x$ is the Milnor number for $h| \tilde{V}_{t_0}$. Since $\tilde{V}_{t_0}$ is a smooth Milnor fiber (of a complete intersection), and φ'_{t_0} is a submersion at x, we can repeat the argument in lemma 5.6 in [DM] to obtain that

$\dim_{\mathbb{C}} D(\tilde{V}_{t_0}, V)_x$ is the length of the Jacobian algebra $\dim_{\mathbb{C}} \mathcal{O}_{\mathbb{C}^{n-k+1},0}/J(h')$ obtained for $h' = h| \tilde{V}_{t_0}$ by choosing local coordinates $(\mathbb{C}^{n-k+1}, 0) \simeq (\tilde{V}_{t_0}, x)$. This is exactly the Milnor number of $h'| \tilde{V}_{t_0}$ at x, completing the proof of the theorem. □

Remark 9.15 : Step 5 of the proof of the theorem for the case of an ICIS V, defined as the transverse intersection of isolated hypersurface singularities, yields exactly the module version of the Lê-Greuel formula. Also, a version of (9.14) is still valid in the case that $\tilde{V}_{t_0}$ is not smooth provided $h| \tilde{V}_{t_0} \setminus Z$ has only isolated singularities in the sense of Lê [Lê1]. However, it is no longer generally true that $\dim_{\mathbb{C}} D(\tilde{V}_{t_0}, V)_x$ is the singular Milnor number of $h| \tilde{V}_{t_0}$ at x (recall example 9.8).

Proof (of proposition 9.14) :

We must establish an analogue of a result of Looijenga [Lo.prop5.4].

Lemma 9.16 : *In the situation of* (9.14), *there exist representatives* $g : X \cap \bar{B}_\varepsilon \to B_{\delta'}$ and $h : X \cap \bar{B}_\varepsilon \to B_\delta$ *such that there is a* δ_1 *with* $0 < \delta_1 < \delta$ *so that:*

i) *the set of points* (t, s) *for which* $g^{-1}(t)$ *and* $h^{-1}(s)$ *are not transverse in* B_ε *is a curve in* $B_{\delta'} \times B_{\delta_1}$ *and*

ii) *there is a homeomorphism* $\varphi : X_1 \to X_2$ *where* $X_1 = X \cap g^{-1}(B_{\delta'}) \cap \bar{B}_\varepsilon$ and $X_2 = X \cap g^{-1}(B_{\delta'}) \cap h^{-1}(B_\delta) \cap \bar{B}_\varepsilon$ *such that* $h \circ \varphi = h$.

The proof of the Lemma will be postponed until we complete the proof of proposition 9.14, which (modulo the lemma) basically follows the line of proof of [Si]. We let $h' = h| X \cap B_\varepsilon$.

Then, we can follow the argument of Siersma [Si]. By our assumption, $h'^{-1}(B_\delta)$ is a trivial fibration off 0 and the finite number of singular values w_j. We repeat Siersma's argument to conclude that $h'^{-1}(B_\delta)$ is obtained from $Z_{0,t_0,\varepsilon}$ $(= X \cap Y_{t_0} \cap Z \cap B_\varepsilon)$ by adjoining (k-1)-cells; and the number of such cells is the sum of the singular Milnor numbers for the singular points of $h'| (X \cap Y_{t_0} \cap B_\varepsilon) \setminus Z$. Since the intersections are topologically trivial over the Milnor sphere S_ε, for s sufficiently small, all singular points of $h| (X \cap Y_{t_0}) \setminus Z$ will be in B_ε. However, by lemma 9.16,

$$(h'^{-1}(B_\delta), Z_{0,t_0,\varepsilon}) \simeq (Y_{t_0,\varepsilon}, Z_{0,t_0,\varepsilon}).$$

Thus, the result follows for the pair $(Y_{t_0,\varepsilon}, Z_{0,t_0,\varepsilon})$. □

Lastly, the proof of lemma 9.16 follows Looijenga [L,prop.5.4]. For $0 < \varepsilon < \varepsilon_0$, X, $X \cap Y$, and $X \cap Y \cap Z$ are (stratified) transverse to S_ε. Next, by assumption, the set of points in $X \setminus Z$ at which dg and dh are linearly dependent form an analytic curve. It maps by (g, h) to an analytic curve α in $\mathbb{C}^2$ which intersects each $\{t\} \times \mathbb{C}$ in a finite number of points. Thus, given $\delta' > 0$, there is a $\delta > 0$ and a δ_1 with $0 < \delta_1 < \delta$ so that in $B_{\delta'} \times \mathbb{C}$, α in fact lies in $B_{\delta'} \times B_{\delta_1}$.

Third, by a curve selection argument as in [L, prop. 5.4], the gradients of $r = \| x \|^2 | X \cap Y$ and $q = | h |^2 | X \cap Y$ do not point in the opposite direction when restricted to any smooth stratum of $X \cap Y \setminus Z$ in a punctured neighborhood of 0.

Then let $W = (X \cap \bar{B}_\varepsilon) \setminus X_2$, where $X_2 = X \cap g^{-1}(B_{\delta'}) \cap h^{-1}(B_\delta) \cap \bar{B}_\varepsilon$. By the compactness of $W \cap Y$, we may find δ and δ' so that the gradients of $r_t = \| x \|^2 | X \cap Y_t$ and $q_t = | h |^2 | X \cap Y_t$ do not point in opposite directions for $| t | < \delta'$. Then, by slightly reducing δ, and following [L, prop. 5.4], we construct a stratified vector field ζ on W which is tangent to $W \cap Y_t$, the fibers of g, and so that $dr_t(\zeta)$ and $dq_t(\zeta) > 0$. Then, arguing as in [L], the integral curves of ζ yield the desired homeomorphism. □

Part IV Topology of Compositions and NonREALizability

In this last part we apply the results of the earlier sections to several problems involving singularities formed by the composition of two germs and bifurcation varieties for deformations.

In § 10 we consider singularities $V'',0 = f_0^{-1}(V'),0$ formed by the pull-backs by germs $f_0: \mathbb{C}^n,0 \longrightarrow \mathbb{C}^p,0$ of almost free divisors and complete intersections. By corollary 4.2 and theorem 2, we can algebraically compute the singular Milnor number of V'' if f_0 is a finite map germ with $n < p$. If $n > p$ then V'' is not, in general, almost free, while if $n = p$ then corollary 4.2 doesn't apply.

With theorem 3 we complement these earlier results by giving a formula for the "vanishing Euler characteristic" of V'' in the case f_0 defines an ICIS with $n \geq p$. As a consequence, we obtain a formula (corollary 10.4) for the vanishing Euler characteristic for the generalized Zariski examples. This formula only involves (singular) Milnor numbers for various objects contrasting with the formulas obtained by Nemethi and Massey-Siersma. In addition we obtain a formula for the singular Milnor number of the pull-back of an almost free divisor when $n = p$ (corollary 10.6).

In § 11 we apply earlier results to the computation of the singular Milnor numbers of bifurcation sets. We obtain an interesting and not fully understood property of the topology of universal bifurcation sets. Even in the universal case there are bifurcation phenomena described by nontrivial vanishing cycles which do not occur over the reals.

§10 A General Composition Formula

We consider the situation expressed in the following diagram of fiber squares.

$$\begin{array}{ccccc} \mathbb{C}^n,0 & \xrightarrow{f_0} & \mathbb{C}^p,0 & \xrightarrow{g_0} & \mathbb{C}^m,0 \\ \uparrow & & \uparrow & & \uparrow i \\ V'',0 & \longrightarrow & V',0 & \longrightarrow & V,0 \end{array}$$

fig. 10.1

We suppose that f_0 defines an isolated complete intersection singularity, $f_0 \mathring{\pitchfork}_{alg} V'$. Also, V is either a free divisor or $= \prod V_i$ where each $V_i \subset \mathbb{C}^{m_i}$, $1 \leq i \leq k$, is a free divisor, and $m = \sum m_i$. Thirdly, we suppose $g_0 \mathring{\pitchfork}_{alg} V$, and hence $V',0 = g_0^{-1}(V),0$ is almost free.

As we remarked earlier, $V'',0$ need no longer be almost free if $n > p$ (however, it remains free up to codimension p). Nonetheless, we are still interested in computing its "vanishing Euler characteristic" and relating it to $\mu(f_0)$ and $\mu_V(g_0)$. For this we consider a stabilization $(f_t, g_{t'})$ of (f_0, g_0), by this we mean: f_t is a stabilization for f_0 and $g_{t'}$ is a stabilization for g_0 so that $\{(t, t'): f_t$ is algebraically transverse to $g_{t'}^{-1}(V)$ in a (fixed) neighborhood U of $0\}$ is a nonempty Zariski open subset. The *vanishing Euler characteristic* is defined to be (where k is the codimension of V)

$$\tilde{\chi}(V'') = (-1)^{n-k}\left(\chi((g_{t'} \circ f_t)^{-1}(V)) - 1\right)$$

where $(f_t, g_{t'})$ is a stabilization, and t and t' are chosen sufficiently small and so that $f_t \circ g_{t'}$ is algebraically transverse to V in a neighborhood of 0. That this is well defined is a consequence of lemma 10.8. If $g_0 \circ f_0$ were algebraically transverse to V off 0 then this would be the singular Milnor number.

We give a formula for this vanishing Euler characteristic in terms of the singular Milnor numbers. We let $\chi(f_0)$ denote the Euler characteristic of the Milnor fiber of f_0, and $\mu_{V'}(f_0)$ and $\mu_V(g_0)$ denote the singular Milnor numbers of f_0 and g_0. We note that provided V and V′ satisfy the conditions of theorem 4.1 or corollary 4.2, these numbers can be computed as $\nu_{V'}(f_0)$ and $\nu_V(g_0)$. However, the next theorem and its proof do not depend upon this

Theorem 3: *In the situation of fig. 10.1 with the above stated assumptions,*

$$\tilde{\chi}(V'') = \mu_{V'}(f_0) + (-1)^{n-p}\mu_V(g_0)\chi(f_0)$$

Remark: The theorem is also valid when there are only geometric stabilizations; this may be necessary as algebraic stabilizations may not exist (see [DM]). That $\tilde{\chi}(V'')$ is still well-defined results from a variant of the argument for lemma 10.8 using [LêT2].

Before giving the proof we indicate two special cases where this formula does give information about the composition $g_0 \circ f_0$.

Example 10.2: *Generalized Zariski Examples* Consider the case

$$\mathbb{C}^n,0 \xrightarrow{f_0} \mathbb{C}^2,0 \xrightarrow{g_0} \mathbb{C},0$$

where g_0 defines an isolated curve singularity $\mathcal{C}$ in $\mathbb{C}^2,0$ and $f_0\colon \mathbb{C}^s,0 \longrightarrow \mathbb{C}^2,0$ defines an isolated complete intersection singularity, with f_0 transverse to $\mathcal{C} = g_0^{-1}(0)$ off 0. Then, for all $0 < |t| < \varepsilon$, $g_0(y)-t$ is a stabilization of g_0; and for almost all $\{(y, t) : g_0(y)=t\}$, $f_0(x)-y$ is a stabilization of f_0; thus, for almost all $0 < |t| < \varepsilon$,

(10.3) $$\tilde{\chi}(f_0^{-1}(\mathcal{C})) = (-1)^{n-1}\left(\chi((g_0 \circ f_0)^{-1}(t)) - 1\right).$$

We denote the RHS of 10.3 by $\tilde{\chi}(f_0 \circ g_0)$, the vanishing Euler characteristic of the non-isolated singularity $(f_0 \circ g_0)^{-1}(0) = f_0^{-1}(\mathcal{C})$. The singular set of $f_0^{-1}(\mathcal{C})$ is $f_0^{-1}(0)$ and along this set it has transverse singularity type $\mathcal{C}$. Nemethi [N] and Massey-Siersma [MS] have given formulas for $\tilde{\chi}(f_0 \circ g_0)$. That of Nemethi involves $\mu(g_0)$, the Euler characteristic of the critical set Σ_t of $f_0 \circ (g_0(y)-t)$, for generic t, and intersection numbers of the components of the critical set and zero set; and in the case n = 3, Massey-Siersma give a formula where the intersection numbers are replaced by a sum of the second Betti numbers of the Local Milnor fibers at points of Σ_t. Using theorem 3, we give an alternate formula only involving Milnor numbers.

Corollary 10.4: *For the generalized Zariski examples in 10.2 ,*

$$\tilde{\chi}(g_0 \circ f_0) = \nu_{\mathcal{C}}(f_0) + (-1)^n \mu(g_0)\chi(f_0)$$

Example 10.5: Consider the classical Zariski example

$$F(x, y) = (x^2 + y^2)^3 + (y^3 + z^3)^2.$$

Here $f_0(x, y, z) = (x^2 + y^2, y^3 + z^3)$ and $g_0(u, v) = u^3 + v^2$ defines a cusp singularity. Then, $\mu(g_0) = 2$ and $\chi(f_0) = (-1)(13) + 1 = -12$. Finally, as

Derlog(g_0) is generated by $2v\frac{\partial}{\partial u} - 3u^2\frac{\partial}{\partial v}$, we compute by proposition 4.3,

$$\nu_{\mathcal{C}}(f_0) = \dim_{\mathbb{C}} \mathcal{O}_{\mathbb{C}^3,0}{}^2/\mathcal{O}_{\mathbb{C}^3,0}((2x, 0), (2y, 3y^2), (0, 3z^2), (2(y^3 + z^3), -3(x^2 + y^2)^2))$$

This is given by the Macaulay-Bezout number (in the notation of [D4])

$$B(\mathbf{d}; \mathbf{a}, \mathbf{c}) \qquad \text{where } \mathbf{a} = (1, 1, 1), \mathbf{c} = (0, 1), \text{ and } \mathbf{d} = (1, 1, 1, 3).$$

The matrix of weight vectors and the degree matrix are given by

$$\begin{pmatrix} 1 & 2 \\ 1 & 2 \\ 1 & 2 \\ 3 & 4 \end{pmatrix} \qquad \mathbf{D} = \begin{pmatrix} 1 & 2 \\ 1 & 2 \\ 1 & 4 \end{pmatrix}$$

Then, $$\tau(\mathbf{D}) = 1 + 4 + 8 + 16 = 29.$$

Thus, applying theorem 2 of [D4], $B(\mathbf{d}; \mathbf{a}, \mathbf{c}) = 29$ and so

$$\tilde{\chi}(F) = 29 + (-1)^3 2(-12) = 53.$$

This is the same result obtained by various other methods.

Next we give a second formula for pullbacks by finite map germs

Pullbacks of almost free divisors by finite map germs

Consider the situation in fig. 10.1, with the same assumptions, where now $n = p$. Since f_0 is a finite map germ, V'' is also an almost free divisor or complete intersection and $\chi(f_0) = \deg(f_0)$. Then, we can use theorem 3 to compute the singular Milnor number for V'' viewed as a section of the almost free divisor V'.

Corollary 10.6: *In the situation of fig. 10.1 with the transversality assumptions and with f_0 a finite map germ with $n = p$,*

$$\mu_{V'}(f_0) = \mu_V(g_0 \circ f_0) - \deg(f_0)\mu_V(g_0).$$

Example 10.7: Consider the pull-back of the cusp $\mathcal{C}$, defined by $g_0(u, v) = u^3 + v^2$, by the finite map germ $f_0(x, y) = (x^2, y^2)$. Then, $\mu(g_0) = 2$, $\deg(f_0) = 4$, and $\mu_V(g_0 \circ f_0) = \mu(g_0 \circ f_0) = 15$ for $g_0 \circ f_0(x, y) = x^6 + y^4$. Thus, by corollary 10.6 $\mu_{\mathcal{C}}(f_0) = 15 - 2 \cdot 4 = 7$. We can also compute this directly since $\mathcal{C}$ is free.

$$\mu_{\mathcal{C}}(f_0) = \nu_{\mathcal{C}}(f_0) = \dim_{\mathbb{C}}(\mathcal{O}_{\mathbb{C}^2,0})^2/\mathcal{O}_{\mathbb{C}^2,0}((2x, 0), (0, 2y), (2y^2, -3x^4))$$

$$= B(\mathbf{d}; \mathbf{a}) \qquad \text{where } \mathbf{a} = (1, 2),\ \mathbf{d} = (1, 2, 4) \text{ (and } \mathbf{c} = (0, 0)),$$

using the notation of [D4].

Again the matrix of weight vectors and the degree matrix are given by

$$\begin{pmatrix} 1 & 1 \\ 2 & 2 \\ 4 & 4 \end{pmatrix} \qquad \mathbf{D} = \begin{pmatrix} 1 & 2 \\ 2 & 4 \end{pmatrix}$$

and so by theorem 2 of [D4]

$$\nu_{\mathcal{C}}(f_0) = (1/2)\tau(\mathbf{D}) = (1/2)(2 + 4 + 8) = 7.$$

We note that a real stabilization of f_0 will have at most $\deg(f_0) = 4$ cusps, no double points, and at most two components. Hence, at most one singular 1-cycle of the 7 complex cycles is REALizable.(i.e. occurs as a vanishing cycle in the real case). In the next section we shall see that this failure of the realization of vanishing cycles for real models can even happen for universal situations.

Before beginning with the proof of the theorem and corollaries, we first establish that the vanishing Euler characteristic is well defined.

Lemma 10.8: *The vanishing Euler characteristic $\tilde{\chi}(V'')$ is well-defined independent of the stabilizations.*

Proof : We give the proof for the case of free divisors. The argument is virtually identical for $V = \prod V_i$ with the V_i free divisors.

Let $(f_{t_1}, g_{t_1'})$ and $(f_{t_2}, g_{t_2'})$ be stabilizations of (f_0, g_0). We first form the sums of these stabilizations

$$\bar{F}(x, t) = f_{t_1} + f_{t_2} - f_0 \quad \text{and} \quad \bar{G}(x, t) = g_{t_1'} + g_{t_2'} - g_0$$

where $t = (t_1, t_2, t_1', t_2') \in T \simeq \mathbb{C}^q$. Also, we let $F(x, t) = (\bar{F}(x, t), t)$, $G(x, t) = (\bar{G}(x, t), t)$ and $\Phi = G \circ F$. We also let $W = V \times T$, and $W' = G^{-1}(W)$.

Next we make several observations regarding the inherited properties of F, G and Φ. First, W is still free; and G is algebraically transverse to W on a neighborhood of $\mathbb{C}^p \times \{0\} \backslash \{(0, 0)\}$ (in fact $G| \mathbb{C}^p \backslash \{0\}$ is by [D2, 1.5 iii)]). Hence, by the proof of prop. 2.12, for $(y, 0) \in W \cap \mathbb{C}^p \times \{0\}$ with $y \neq 0$, $T_{\log}W'_{(y, 0)} = T_{\log}V'_{(y)} \oplus T'$ where T' projects isomorphically onto T. Then, we see that F is algebraically transverse to W' on a neighborhood of $\mathbb{C}^n \times \{0\} \backslash \{(0, 0)\}$, again because $F| \mathbb{C}^n \backslash \{0\}$ is.

Now we choose open neighborhoods of 0, $U \subset \mathbb{C}^m \times T$, $U' \subset \mathbb{C}^p \times T$, and $U'' \subset \mathbb{C}^n \times T$, with $F(U'') \subset U'$ and $G(U') \subset U$. Furthermore, we may assume that a set of generators $\{\zeta_i\}$ of Derlog(W) are defined on U and (by coherence) also generate Derlog(W,z) for all $z \in U$. Also, we may asume both U' and U'' are of the form $U' = U_1 \times U_3$ and $U'' = U_2 \times U_3$ for U_3 a neighborhood of 0 in T.

Lastly we define in U′ and U″

$$Z' = \{ (y, t) \in U' \colon G \text{ is not algebraically transverse to } W \text{ at } (y, t)\} \quad \text{and}$$

$$Z'' = \{ (x, t) \in U'' \colon F \text{ is not algebraically transverse to } W' \text{ at } (x, t)\}.$$

Because algebraic transversality is given by an algebraic condition, both Z′ and Z″ are Zariski closed subsets of U′, respectively U″. For example,

$$Z' = \operatorname{supp}\Big(\mathcal{O}_{\mathbb{C}^{p+q},0} \{\frac{\partial}{\partial z_i}\} / \mathcal{O}_{\mathbb{C}^{p+q},0} \{ \zeta_i \circ G, \frac{\partial \bar{G}}{\partial y_i} \} \Big)$$

and similarly for Z″ using F and Derlog(W′) instead.

Next, consider the projections $\pi' \colon U' \longrightarrow U_3$ and $\pi'' \colon U'' \longrightarrow U_3$ of Z′ and Z″. By our assumptions about the stabilizations, after possibly shrinking U′ and U″, we conclude the complements of $\pi'(Z')$ and $\pi''(Z'')$ are nonempty. Furthermore, $\pi'^{-1}(0) \cap Z'$ and $\pi''^{-1}(0) \cap Z''$ both equal {0}; thus the restrictions $\pi'| Z'$ and $\pi''| Z''$ are both finite mappings; hence, $\pi'(Z')$ and $\pi''(Z'')$ are Zariski closed subsets of U_3. Thus, $\mathcal{U} = U_3 \setminus (\pi'(Z') \cup \pi''(Z''))$ is a nonempty Zariski open subset of U_3.

Note that if $t \in \mathcal{U}$ then G is algebraically transverse to W on $U_2 \times \{t\}$. Similarly, F is algebraically transverse to W′ on $U_2 \times \{t\}$. Hence, by (the proof of) Cor. 2.15, Φ is algebraically (and hence geometrically) transverse to W on $U_2 \times \{t\}$. Lastly,by a standard application of Thom's first isotopy lemma (see e.g. [Ma1, Ma2, Ma3] or [GW]) we conclude that any $t \in \mathcal{U}$ has a neighborhood B on which the topological type of $\Phi^{-1}(W) \cap U_2$ is constant (again it may be necessary to shrink U_2 and U_3 further so that $\Phi| \partial U_2$ is also geometrically transverse to W). Since $\mathcal{U}$ is connected, the topological type is constant on $\mathcal{U}$, completing the proof.

If we weaken the conditions by only requiring geometric stabilizations, then the argument goes through provided we can still conclude that $\mathcal{U}$ is Zariski open, or equivalently, that Z′ and Z″ are Zariski closed. This now follows by applying Theorem 2.3.2 of [LêT2] since by the graph trick we can assume we have embeddings and by a nonlinear change of coordinates a linear subspace. □

Proof (of Theorem 3):

Let $V'_{t'} = g_{t'}^{-1}(V)$ be a stabilization of g_0, defined in an ε_1-neighborhood of 0. We may also assume that $V' \pitchfork_{geom} S_\varepsilon = \partial B_\varepsilon$ for $0 < \varepsilon < \varepsilon_1$. By lemma 10.8, we can use any stabilization f_t of f_0. We may assume (by composing with families of translations if necessary) that $0 \notin D(f_t)$ and $D(f_t) \pitchfork V'$ for $t \neq 0$.

fig. 10.2

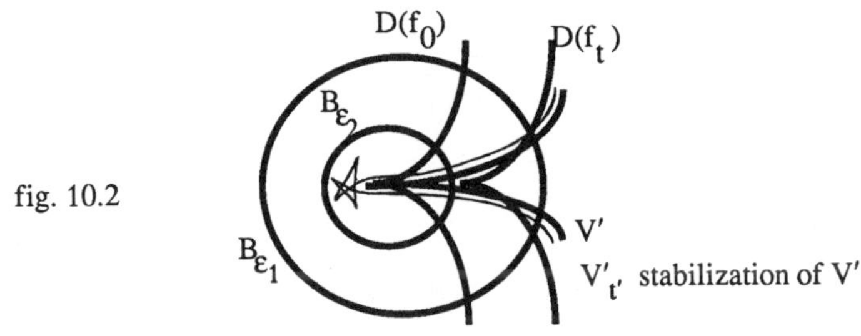

To simplify notation we shall let $\chi_\delta(A)$ denote $\chi(A \cap B_\delta)$ for $A \subset \mathbb{C}^n$ and similarly let $\chi_\varepsilon(A)$ denote $\chi(A \cap B_\varepsilon)$ for $A \subset \mathbb{C}^p$, while $\chi_{\delta,\varepsilon}(f_t^{-1}(A))$ will denote $\chi_\delta(f_t^{-1}(A \cap B_\varepsilon))$ for $A \subset \mathbb{C}^p$. Thus,

$$\tilde{\chi}(V'') = (-1)^{n-k}\left(\chi_{\delta,\varepsilon}(f_t^{-1}(V'_{t'})) - 1\right); \quad \text{where } 0 < \varepsilon << \delta.$$

Now, given t we can pick an $\varepsilon_2 < \varepsilon_1$ so that $B_{\varepsilon_2} \cap D(f_t) = \emptyset$; and we choose γ small enough so that $V'_{t'} \pitchfork_{geom} S_\varepsilon$ for $|t'| < \gamma$ and $\varepsilon_2 \le \varepsilon \le \varepsilon_1$. Now we can compute

$$\textbf{(10.9)} \qquad \chi_{\delta,\varepsilon_1}(f_t^{-1}(V'_{t'})) = \chi_{\delta,\varepsilon_2}(f_t^{-1}(V'_{t'})) + \chi_\delta(f_t^{-1}(V'_{t'} \cap (B_{\varepsilon_1} \backslash B_{\varepsilon_2}))).$$

Then, as $f_t \mid f_t^{-1}(B_{\varepsilon_2}) \cap B_\delta$ is a trivial fibration with fiber the Milnor fiber of f_t, which is also the Milnor fiber of f_0,

$$\begin{aligned} \chi_{\delta,\varepsilon_2}(f_t^{-1}(V'_{t'})) - \chi_{\delta,\varepsilon_2}(f_t^{-1}(V')) &= \left(\chi_{\varepsilon_2}(V'_{t'}) - \chi_{\varepsilon_2}(V')\right) \cdot \chi_\delta(f_t^{-1}(0)) \\ &= \tilde{\chi}(V') \cdot \chi(f_0) \\ \textbf{(10.10)} \qquad &= (-1)^{p-k}\, \mu_V(g_0) \cdot \chi(f_0) \end{aligned}$$

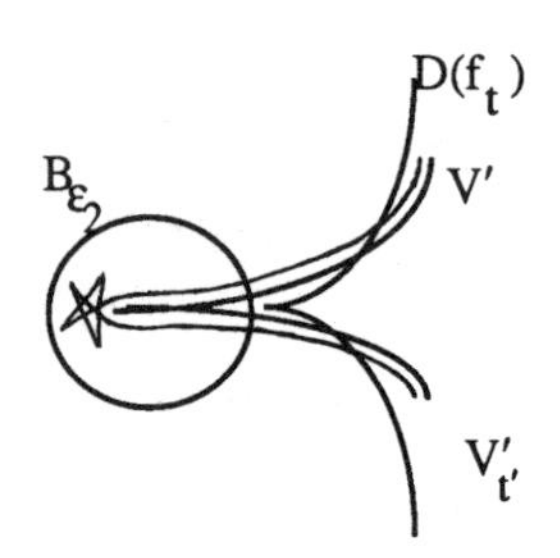

fig. 10.11

Hence, from (10.10)

$$\chi_{\delta,\varepsilon_2}(f_t^{-1}(V'_{t'})) \quad = \quad (-1)^{p-k}\,\mu_V(g_0)\cdot\chi(f_0) \;+\; \chi_{\delta,\varepsilon_2}(f_t^{-1}(V'))$$

and from (10.9)

(10.12)
$$\begin{aligned}\chi_{\delta,\varepsilon_1}(f_t^{-1}(V'_{t'})) \quad &= \quad (-1)^{p-k}\,\mu_V(g_0)\cdot\chi(f_0) + \chi_{\delta,\varepsilon_2}(f_t^{-1}(V')) \\ &\qquad + \chi_\delta(f_t^{-1}(V'_{t'} \cap (B_{\varepsilon_1}\backslash B_{\varepsilon_2}))).\end{aligned}$$

Then, by the geometric transversality of $\bar{B}_{\varepsilon_1}\backslash B_{\varepsilon_2}$ to $V'_{t'}$ for $|t'| < \gamma$, by Thom's first isotopy lemma (see fig 10.13)

$$f_t^{-1}(V' \cap (B_{\varepsilon_1}\backslash B_{\varepsilon_2}))\cap B_\delta \;\simeq\; f_t^{-1}(V'_{t'}\cap (B_{\varepsilon_1}\backslash B_{\varepsilon_2}))\cap B_\delta,$$

and hence,

(10.14)
$$\chi_{\delta,\varepsilon_1}(f_t^{-1}(V')) \quad = \quad \chi_{\delta,\varepsilon_2}(f_t^{-1}(V')) + \chi_\delta(f_t^{-1}(V'_{t'} \cap (B_{\varepsilon_1}\backslash B_{\varepsilon_2}))).$$

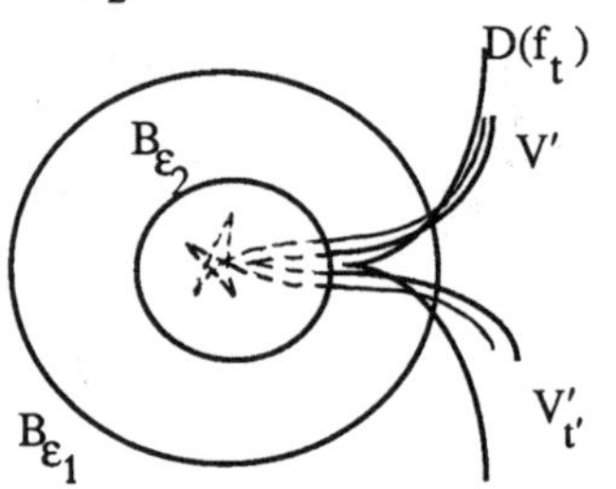

fig. 10.13

Thus, by (10.12) and (10.14)

(10.15)
$$\chi_{\delta,\varepsilon_1}(f_t^{-1}(V'_{t'})) \quad = \quad (-1)^{p-k}\,\mu_V(g_0)\cdot\chi(f_0) + \chi_{\delta,\varepsilon_1}(f_t^{-1}(V'))$$

or

$$\begin{aligned}\tilde{\chi}(V'') \quad &= \quad (-1)^{n-k}\Big(\chi_{\delta,\varepsilon_1}(f_t^{-1}(V'_{t'})) - 1\Big) \\ &= \quad (-1)^{n-p}\,\mu_V(g_0)\cdot\chi(f_0) + (-1)^{n-k}\Big(\chi_{\delta,\varepsilon_1}(f_t^{-1}(V')) - 1\Big) \\ &= \quad (-1)^{n-p}\,\mu_V(g_0)\cdot\chi(f_0) + \mu_{V'}(f_0)\,. \qquad \square\end{aligned}$$

Proof (of the Corollaries):

For corollary 10.4 we apply theorem 3 with $V = \{0\}$ and g_0 defining the isolated curve singularity $C = V'$. Then, $\mu_V(g_0) = \mu(g_0)$ and $\mu_{V'}(f_0) = \nu_C(f_0)$. Thus, it remains to show

(10.16) $$\tilde{\chi}(V'') = \tilde{\chi}(g_0 \circ f_0).$$

For this we claim that $(f_0, g_{t'})$ is a stabilization, where $g_{t'} = g_0(y) - t'$, so that $V'_{t'}$ is the Milnor fiber of g_0. This follows because by the parametrized transversality theorem f_0 is transverse to the Milnor fibers $g_0^{-1}(t')$ for almost all t'. Then,

$$\tilde{\chi}(V'') = (-1)^{n-1}\left(\chi((g_{t'} \circ f_0)^{-1}(0)) - 1\right);$$

and for $0 < |t'| < \varepsilon << \delta$,

$$\chi((g_{t'} \circ f_0)^{-1}(0)) = \chi_\delta((g_0 \circ f_0)^{-1}(t')) = \tilde{\chi}(g_0 \circ f_0). \quad \square$$

For corollary 10.6, $g_0 \circ f_0$ is algebraically transverse to V off 0 so V'' is almost free and $\tilde{\chi}(V'') = \mu_V(g_0 \circ f_0)$. As f_0 is a 0-dimensional ICIS, $\chi(f_0) = \deg(f_0)$ and the result follows. $\square$

§11 Topology of Bifurcation Sets and NonREALizability

In this section we consider vanishing cycles for bifurcation sets and see how they relate to the nonREALizability question.

NonREALizability Question : Suppose $V \subset \mathbb{C}^n$ is an almost free divisor or complete intersection based on $V' \subset \mathbb{C}^p$. V has *a real form* if $V' = V''_{\mathbb{C}}$ for $V'' \subset \mathbb{R}^n$ and there is a germ $f_0 : \mathbb{R}^n,0 \to \mathbb{R}^p,0$ whose complexification defines an almost free V_1 $\mathcal{K}_V$-equivalent to V. Then, V is REALizable if it has a real form V_1 with a real stabilization having $\mu(V)$ number of vanishing cycles. Thus, all of the vanishing cycles can be realized in some real representative.

Some of the examples from the preceding sections illustrate that not all almost free V are realizable. However, in certain universal situations we do have REALizability. For finite stable map germs $f : \mathbb{R}^n,0 \to \mathbb{R}^p,0$, so $n \leq p$, we can define the local multiplicity m(f) of f as the maximum number of solutions to f(x) = y which can occur for y and x sufficiently close to 0 (see [DG]). Surprisingly, by [DG] (and a remark in [DG2]), for a wide range of stable germs f, both $m(f) = m(f_{\mathbb{C}}) = \dim_{\mathbb{R}} Q(f)$, for Q(f) the local algebra of f. Thus, the maximum number of complex solutions are realized over $\mathbb{R}$ for stable germs.

We might hope that vanishing cycles are likewise REALizable for universal situations such as discriminants, bifurcation sets etc. We shall see that this is not the case for bifurcation sets.

Consider $f_0 : \mathbb{C}^n,0 \longrightarrow \mathbb{C},0$ defining an isolated hypersurface singularity; and let $F : \mathbb{C}^{n+q},0 \longrightarrow \mathbb{C}^{1+q},0$ denote its miniversal unfolding for $\mathcal{R}^+$-equivalence. Using local coordinates **x** for $\mathbb{C}^n$, y for $\mathbb{C}$, and **u** for $\mathbb{C}^q$, we represent $F(x, u) = (\bar{F}(x, u), u)$. Then, we may assume $\bar{F}(x, 0) \equiv 0$. Then being miniversal means

$$\{1, \frac{\partial \bar{F}}{\partial u_1}\Big|_{u=0}, \ldots , \frac{\partial \bar{F}}{\partial u_q}\Big|_{u=0}\} \text{ forms a basis for } \mathcal{O}_{\mathbb{C}^n,0} / (\frac{\partial f_0}{\partial x_1}, \ldots , \frac{\partial f_0}{\partial x_n}) .$$

We let D(F) denote the discriminant of F. We let $\pi : \mathbb{C}^{1+q},0 \longrightarrow \mathbb{C}^q,0$ denote projection; and if h is a defining equation for D(F), written as a Weierstrass polynomial in y, then (see e.g. [To2])

$$\mathrm{Sing}(D(F)) = \{(y, u) : h(y, u) = \frac{\partial h}{\partial y}(y, u) = 0\}$$

The bifurcation set of F, $B(F) \subset \mathbb{C}^q$, is defined to be $\pi(\mathrm{Sing}(D(F)))$. It is a free divisor by [To2] and [Br]. By [Thm B,To2] and [Thm 5,Br], Derlog(B(F)) consists of ζ which have lifts to $\eta \in \mathrm{Derlog}(D(F))$ which are then of the form $\eta = g(y, u)\frac{\partial}{\partial y} + \zeta$. Then, Derlog(B(F)) is freely generated by those ζ_i which have lifts $\eta_i = y^i \cdot \frac{\partial}{\partial y} + \zeta_i$, $1 \leq i \leq q$. However, as $\eta \in \mathrm{Derlog}(D(F))$, by [A] and [Sa] it is liftable via F to ξ so $\xi(F) = \eta \circ F$. We denote the lift of η_i by ξ_i. Then,

$$(\xi, \eta) \in T\mathcal{A}_{eq,e} = \mathcal{O}_{\mathbb{C}^{n+q},0}\{\frac{\partial}{\partial x_i}\} \oplus \mathcal{O}_{\mathbb{C}^{1+q},0}\{\frac{\partial}{\partial y}\} \oplus \mathcal{O}_{\mathbb{C}^q,0}\{\frac{\partial}{\partial u_i}\}$$

where $\mathcal{A}_{eq}$ denotes the group of $\mathcal{A}$-equivalence of unfoldings. It consists of pairs of diffeomorphisms $\varphi : \mathbb{C}^{n+q},0 \longrightarrow \mathbb{C}^{n+q},0$ and $\psi : \mathbb{C}^{1+q},0 \longrightarrow \mathbb{C}^{1+q},0$ of the form $\varphi(x, u) = (\varphi_1(x, u), \chi(u))$ and $\psi(x, u) = (\psi_1(x, u), \chi(u))$ for a germ of a diffeomorphism $\chi : \mathbb{C}^q,0 \longrightarrow \mathbb{C}^q,0$. It is the composition of an element of the unfolding group $\mathcal{A}_{un}$ which fixes u with a diffeomorphism χ of $\mathbb{C}^q,0$. It follows that if we let (φ_t, ψ_t) denote the flow generated by (ξ, η) and χ_t that for ζ, then (φ_t, ψ_t) gives a trivialization of F under $\mathcal{A}_{eq}$, and χ_t gives a trivialization of B(F). Thus, although B(F) is usually defined as above using $\mathcal{R}^+$-equivalence, *it is really a property of $\mathcal{A}$-equivalence of unfoldings.*

Now let $f : \mathbb{C}^{n+r},0 \longrightarrow \mathbb{C}^{1+r},0$ be an unfolding f_0 of the form $f(x, v) = (\bar{f}(x, v), v)$, where we may assume (by applying the equivalence $y \mapsto y - \bar{f}(0, v)$) that $\bar{f}(0, v) \equiv 0$. By the versality of F, there is a germ $\lambda : \mathbb{C}^r,0 \longrightarrow \mathbb{C}^q,0$ such that λ^*F is $\mathcal{A}_{un}$-equivalent to f. We establish an analogue of [prop. 2.3, D2] for bifurcation. For it we define

$$\mathcal{A}_{eq,e}\text{-codim}(f) \stackrel{\text{def}}{=} \dim_{\mathbb{C}}\Big(\mathcal{O}_{\mathbb{C}^{n+q},0}\{\tfrac{\partial}{\partial y}\}/\Big(\mathcal{O}_{\mathbb{C}^{n+q},0}\{\tfrac{\partial \bar{f}}{\partial x_i}\}+\mathcal{O}_{\mathbb{C}^{1+q},0}\{\tfrac{\partial}{\partial y}\}+\mathcal{O}_{\mathbb{C}^{q},0}\{\tfrac{\partial \bar{f}}{\partial v_i}\}\Big)\Big).$$

Definition 11.1 : The unfolding f has *finite bifurcation codimension* if

$$\mathcal{A}_{eq,e}\text{-codim}(f) \;<\; \infty$$

Then we have

Proposition 11.2 : *The unfolding* f *has finite bifurcation codimension iff* λ *has finite* $\mathcal{K}_{B(F)}$*-codimension, i.e.* $\lambda \mathrel{\mathring{\pitchfork}}_{\text{alg}} B(F)$.

Before proving this proposition, we use it to compute the number of vanishing cycles for an unfolding of finite bifurcation codimension. We have indicated that by the results of Bruce and Terao, Derlog(B(F)) is generated by the $\{\zeta_i\}$. By replacing F by $F \times id_{\mathbb{C}}$ we may assume that B(F) has a good defining equation H and that $\{\zeta_1, \dots, \zeta_{q-1}\}$ generate Derlog(H). For example, if f_0 is weighted homogeneous, then we can also assume that F and H are; then there is an Euler vector field, and subtracting an appropriate multiple gives the generators of Derlog(H).

Corollary 11.3 : *For a finite bifurcation germ* f *induced from* F *by* $\lambda : \mathbb{C}^r,0 \to \mathbb{C}^q,0$, *suppose that* B(f) *is finitely defined via* λ. *The number of vanishing cycles* is given by

$$\nu_{B(F)}(\lambda) \;=\; \dim_{\mathbb{C}}\left(\mathcal{O}_{\mathbb{C}^r,0}\right)^q/\mathcal{O}_{\mathbb{C}^r,0}\{\tfrac{\partial \lambda}{\partial v_i}, \zeta_j\circ\lambda, 1 \le j \le q-1\}$$

Example 11.4 : Consider the versal unfolding of the A_4 singularity

$$F(x, u) = (x^5 + u_3 \cdot x^3 + u_2 \cdot x^2 + u_1 \cdot x\,, u_1, u_2, u_3\,)$$

We consider the two dimensional section $\lambda : \mathbb{C}^2,0 \to \mathbb{C}^3,0$ given by $\lambda(v_1, v_2) =$

$(v_1, v_2, 0)$ and the induced unfolding $f(x, v) = (x^5 + v_2 \cdot x^2 + v_1 \cdot x, v_1, v_2)$.
We note that F as well as the section are weighted homogeneous for the same weights $wt(x, y, u_1, u_2, u_3, v_1, v_2) = (1, 5, 4, 3, 2, 4, 3)$. Hence we can determine the weights of generators for Derlog(H) without explicitly calculating them. By the construction of Terao–Bruce the ζ_i will be weighted homogeneous and lift to $\eta_i = y^i \cdot \frac{\partial}{\partial y} + \zeta_i$, with $i > 1$ for the non-Euler vector fields. Thus, $wt(\zeta_i) = 5$ for $i = 2$ and 10 for $i = 3$. Hence, we can compute $\nu_{B(F)}(\lambda)$ as a weighted Macaulay–Bezout number. If $g_i(v)$ denotes the $\frac{\partial}{\partial u_3}$–component of ζ_i, then $wt(g_i(v)) = 5+2 = 7$ for $i = 2$ and $10+2 = 12$ for $i = 3$. Thus,

$$\nu_{B(F)}(\lambda) = \dim_{\mathbb{C}} \mathcal{O}_{\mathbb{C}^2,0}/(g_2(v), g_3(v)) = 7 \cdot 12/4 \cdot 3 = 7.$$

Thus, there are 7 vanishing cycles for B(f). This is in contrast with the number found earlier for the discriminant which was 1 since f has $\mathcal{A}_e$–codim(f) = 1 from §4. In fact, this is the generic section to B(F). However, the maximum number of vanishing cycles for a real section is 5. They correspond to the number of compact regions bounded by a perturbation of a section through 0 as in fig. 11.5.

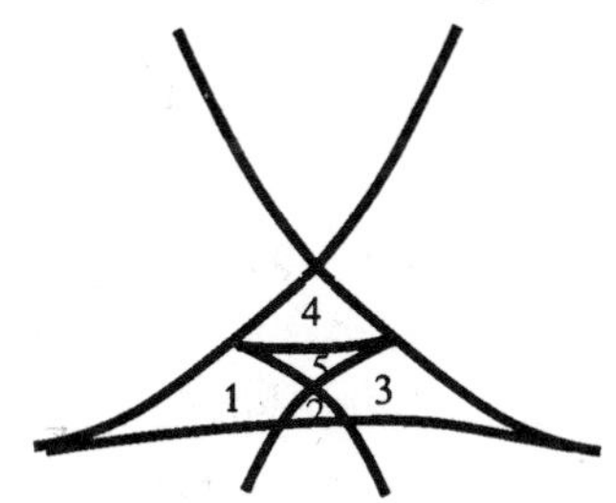

fig. 11.5

The regions correspond to the number of different morsifications of x^5 having 4 real critical points. There are 5 such. Arnold has discovered general formulas for the number of such Morsifications and related them to Springer numbers [A2]. Since this is the maximum number for real Morsifications, there are

2 vanishing cycles which are not realizable! It remains to be seen what complex phenomena these cycles are capturing. It would also be interesting to determine the number of vanishing cycles for higher A_n and relate them to Arnold's results.

We now turn to the proof of proposition 11.2. We will prove it by following the line of reasoning given in [D2, §2]. For it we need a geometric characterization of finite $\mathcal{A}_{eq}$-determinacy. We let, as usual, $\Sigma(f)$ denote the critical set of f. Also, we choose neighborhoods U_1 and U_2 of 0 in $\mathbb{C}^n$ and $\mathbb{C}^r$ and $U = U_1 \times U_2$ in $\mathbb{C}^{n+r}$ and W of 0 in $\mathbb{C}$ and a representative of f on U such that:

i) $f^{-1}(0) \cap \Sigma(f) = \{0\}$;

ii) $f \mid \Sigma(f) : \Sigma(f) \to W \times U_2$ is finite to one;

iii) $\Sigma(f) \cap (U_1 \times \{0\}) = \{0\}$

iv) the projection $\pi' : \Sigma(f) \to \mathbb{C}^r$ is finite to one (this is just the catastrophe map).

Then, for $u_0 \in U_2$ we let $S = \Sigma(f) \cap U_1 \times \{u_0\}$. Then, (F, S) defines an unfolding of a multi-germ $(\bar{f}(\cdot, u_0), S)$. The versality theorem holds for $\mathcal{A}$-equivalence of multi-germs (for example by applying the versality theorem in [D5]).

Lemma 11.6 : *The unfolding* f *has finite bifurcation codimension (i.e. finite* $\mathcal{A}_{eq}$*-codimension) iff for all* u_0 *in a punctured neighborhood of 0,* (F, S) *is an* $\mathcal{A}$*-versal unfolding of* $(\bar{f}(\cdot, u_0), S)$.

Proof (of lemma 11.6) : The proof is very similar to the corresponding Mather-Gaffney geometric characterization of finite $\mathcal{A}$-codimension using the coherence of sheaves and Grauert's theorem on direct images. We let $\Theta(\bar{f})$ denote the sheaf of vector fields along $\bar{f}$ on U, and Θ_n the sheaf of vector fields on U. Both of these are coherent. Then, $t\bar{f} : \Theta_n \to \Theta(\bar{f})$ is defined by $t\bar{f}(\xi) = \xi(\bar{f})$; and, $\Theta(\bar{f})/t\bar{f}(\Theta_n)$ is coherent with support on $\Sigma(f)$. Since $f : \Sigma(f) \to W \times U_2$ is finite to one, Grauert's theorem on direct images of coherent sheaves implies $\mathcal{N}_1(\bar{f}) = f_*(\Theta(\bar{f})/t\bar{f}(\Theta_n))$ is a coherent sheaf of $\mathcal{O}_{W \times U_2}$-modules (with $\mathcal{O}_{W \times U_2}$ denoting the sheaf of holomorphic functions on $W \times U_2$).

Second, we define $\omega\bar{f} : \Theta(\pi_1) \to \mathcal{N}_1(\bar{f})$ where $\pi_1 : \mathbb{C}^{1+r} \to \mathbb{C}$ denotes projection; $\omega\bar{f}(\eta) = \eta\circ f$. Again the cokernel $\mathcal{N}_1(\bar{f})/\omega\bar{f}(\Theta(\pi_1))$ is coherent with support on D(f). Moreover the restriction of the projection $\pi : \mathbb{C}^{1+r} \longrightarrow \mathbb{C}^r$ to D(f) is a finite to one mapping because its composition with f is $\pi' : \sum(f) \longrightarrow \mathbb{C}^r$ which is a finite to one by iv) above. Thus, again Grauert's theorem implies $\pi_*(\mathcal{N}_1(\bar{f})/\omega\bar{f}(\Theta(\pi_1)))$ is a coherent sheave of $\mathcal{O}_{U_2}$-modules.

Lastly, we define $\rho : \Theta_r \longrightarrow \pi_*(\mathcal{N}_1(\bar{f})/\omega\bar{f}(\Theta(\pi_1)))$ by $\zeta \mapsto \zeta(\bar{f})$. and let $\mathcal{N}(\bar{f})$ denote the cokernel. Then for $u_0 \in U_2$, the stalk $\mathcal{N}(\bar{f})_{u_0} = 0$ iff for

$$S = \{(x_1, u_0), \dots, (x_s, u_0)\} = \pi'^{-1}(u_0) \cap \textstyle\sum(f) \quad \text{and}$$
$$S' = \{(y_1, u_0), \dots, (y_m, u_0)\} = f(S)$$

$$\text{(11.7)} \quad \oplus\, \theta(\bar{f}, (x_i, u_0)) = \sum t\bar{f}\, \theta_{n(x_i, u_0)} + \sum \omega\bar{f}\, \theta(\pi_1)_{(y_i, u_0)} + \rho(\theta_{r\,(u_0)})$$

where $\rho(\zeta) = (\zeta(\bar{f})_{(x_1, u_0)}, \dots, \zeta(\bar{f})_{(x_s, u_0)})$ and if $\eta \in \theta(\pi_1)_{(y_i, u_0)}$ then $\omega\bar{f}(\eta) = (\eta\circ f_{(x_{i_1}, u_0)}, \dots, \eta\circ f_{(x_{i_k}, u_0)})$ for those entries where $f(x_{i_j}, u_0) = (y_i, u_0)$.

Then, (11.7) is exactly the infinitesimal condition that the unfolding f of the multi-germ $\bar{f}(\cdot, u_0)$: $\mathbb{C}^n, S \longrightarrow \mathbb{C}, S'$ be $\mathcal{A}$-versal. Thus, f is an $\mathcal{A}$-versal unfolding of $\bar{f}(\cdot, u_0)$: $\mathbb{C}^n, S \longrightarrow \mathbb{C}, S'$ for all u_0 in a punctured neighborhood of 0 implies $\mathrm{supp}(\mathcal{N}(\bar{f})) = \{0\}$. Then, by the Nullstellensatz for coherent analytic sheaves (see e.g. [Tg, chap.2,§7]), there is a k such that $m^k_{\mathbb{C}^q,0}\cdot\mathcal{N}(\bar{f})_0 = 0$. However, by coherence of $\mathcal{N}(\bar{f})$, $\mathcal{N}(\bar{f})_0$ is a finitely generated $\mathcal{O}_{\mathbb{C}^q,0}$-module, Nakayama's lemma implies by $\dim_{\mathbb{C}}\mathcal{N}(\bar{f})_0 < \infty$. However,

$$\mathcal{N}(\bar{f})_0 = \mathcal{O}_{\mathbb{C}^{n+q},0}\{\tfrac{\partial}{\partial y}\}/\Big(\mathcal{O}_{\mathbb{C}^{n+q},0}\{\tfrac{\partial \bar{f}}{\partial x_i}\} + \mathcal{O}_{\mathbb{C}^{1+q},0}\{\tfrac{\partial}{\partial y}\} + \mathcal{O}_{\mathbb{C}^q,0}\{\tfrac{\partial \bar{f}}{\partial u_i}\}\Big).$$

For the converse, we reverse the argument at the last stage to observe that finite bifurcation codimension implies $\dim_{\mathbb{C}}\mathcal{N}(\bar{f})_0 < \infty$ or for k sufficiently large $m^k_{\mathbb{C}^q,0}\cdot\mathcal{N}(\bar{f})_0 = 0$. Then, by the coherence $\mathcal{N}(\bar{f})$ there are a set $\{\varphi_i\}$ of generators $\mathcal{N}(\bar{f})_0$ which generate $\mathcal{N}(\bar{f})$ in a neighborhood of 0. Then, $u^\alpha\cdot\varphi_i = 0$ for $|\alpha| \geq k$. Thus, at any point $u_0 \neq 0$ there is a monomial $u^\alpha \neq$ at u_0. Thus, unit$\cdot\varphi_i = 0$ at u_0

or $\mathcal{N}(\bar{f})_{u_0} = 0$. Thus, f is an $\mathcal{A}$-versal unfolding of $\bar{f}(\cdot,u_0)$: $\mathbb{C}^n,S \longrightarrow \mathbb{C}, S'$ for all u_0 in a punctured neighborhood of 0. □

Proof (of proposition 11.2) : For both directions we use the geometric criterion from the preceding lemma.

By shrinking U_1, U_2, and W if necessary we may assume :

1) by replacing F by $F \times id_{\mathbb{C}^r}$, we still have $B(F \times id_{\mathbb{C}^r}) = B(F) \times \mathbb{C}^r$ free and we may assume that λ is an embedding;

2) the generators $\{\zeta_i\}$ are defined on U_2 and generate the associated sheaf Derlog(B(F)) at all $u \in U_2$;

3) the lifts $\{\eta_i\}$ are defined on $W \times U_2$ and generate the lifts of Derlog(B(F)) at points of $W \times U_2$;

4) the lifts $\{\xi_i\}$ of the $\{\eta_i\}$ are defined on U.

$\Leftarrow$ As λ is finitely $\mathcal{K}_{B(F)}$-determined it is transverse to B(F) in a punctured neighborhood of 0. We may also assume (by shrinking everything further if necessary)

5) λ is a representative which is algebraically transverse to Derlog(B(F)) on $U_2 \cap \mathbb{C}^r \setminus \{0\}$.

For $u_0 \in U_2 \cap \mathbb{C}^r \setminus \{0\}$, let $S = \pi'^{-1}(u_0) \cap \Sigma(f)$, which is finite. Then, F : $\mathbb{C}^{n+q},S \rightarrow \mathbb{C}^{1+q},S'$ is an $\mathcal{A}$-versal unfolding of the multi-germ $\bar{f}(\cdot,u_0)$: $\mathbb{C}^n,S \longrightarrow \mathbb{C},S'$. Pick a subset $\{\zeta_1, \ldots ,\zeta_m\}$ of the above set $\{\zeta_i\}$ such that $<\zeta_{1(\lambda(u_0))}, \ldots ,\zeta_{m(\lambda(u_0))}>$ spans a complementary subspace to $d\lambda(u_0)(T\mathbb{C}^r)$. Then, since $\xi_i(F) = F \circ \eta_i$, by a standard argument as e.g. in Martinet [Ma1], F : $\mathbb{C}^{n+q},S \rightarrow \mathbb{C}^{1+q},S'$ is $\mathcal{A}$-equivalent as an unfolding of a multi-germ to $f \times id$: $\mathbb{C}^{n+q},S \rightarrow \mathbb{C}^{1+q},S'$. This implies that $f : \mathbb{C}^{n+r},S \rightarrow \mathbb{C}^{1+r},S'$ is $\mathcal{A}$-versal (f is $\mathcal{A}$-versal iff $f \times id$ is by the infinitesimal criteria for versality,see [D5]). As u_0 was an arbitrary point of $U_2 \cap \mathbb{C}^r \setminus \{0\}$, f is is an $\mathcal{A}$-versal unfolding of the multi-germ $\bar{f}(\cdot,u_0)$ for all u_0 in a punctured neighborhood of 0. By Lemma 11.5, f has finite bifurcation codimension.

$\Rightarrow$ Conversely, if f has finite bifurcation codimension, then for u_0 in a punctured neighborhood $U_2 \cap \mathbb{C}^r \setminus \{0\}$, $f : \mathbb{C}^{n+r}, S \to \mathbb{C}^{1+r}, S'$ is $\mathcal{A}$-versal. Hence, $F : \mathbb{C}^{n+q}, S \to \mathbb{C}^{1+q}, S'$ is an $\mathcal{A}$-trivial extension of f (i.e. $\mathcal{A}$-equivalent as an unfolding to $f \times id$). Thus, there are vector fields ξ'_i, η'_i defined near S and S′ so that $\xi'_i(F) = \eta'_i \circ F$, both ξ_i and η'_i lift ζ'_i, and $\{\zeta'_{i(\lambda(u_0))}\}$ span a subspace complementary to $d\lambda(u_0)(T\mathbb{C}^r)$. Thus, by the characterization of Bruce and Terao, $\zeta'_i \in \mathrm{Derlog}(B(F))_{\lambda(u_0)}$. By 2), $\{\zeta_i\}$ generate $\mathrm{Derlog}(B(F))_{\lambda(u_0)}$. Hence, the subspace spanned by $\{\zeta'_{i(\lambda(u_0))}\}$ is contained in that spanned by $\{\zeta_{i(\lambda(u_0))}\}$. Thus, $\lambda(\mathbb{C}^r)$ is algebraically transverse to $\mathrm{Derlog}(B(F))$ at $\lambda(u_0)$. Thus, $\lambda(\mathbb{C}^r)$ is algebraically transverse to $\mathrm{Derlog}(B(F))$ in the punctured neighborhood of 0, i.e. λ is finitely $\mathcal{K}_{B(F)}$-determined. □

Bibliography

A Arnold, V.I. *Normal forms of functions near degenerate critical points, Weyl groups* A_k, D_k, E_k *and Lagrange singularities,* Funct. Anal. and Appl. 6 (1972) 254-272

A2 ------ *Springer Numbers and Morsification Spaces,* Jour. Alg. Geom. 1 (1992) 197-214

BR Bruce, J.W. and Roberts, M. *Critical points of functions on analytic varieties,* Topology vol. 27 no. 1, 1988, 57-91

Br Bruce, J.W. *Vector fields on discriminants and Bifurcation Varieties,* Bull. London Math. Soc. 17 (1985) 257-262

D1 Damon, J. *Deformations of sections of singularities and Gorenstein surface singularities,* Amer. J. Math. 109 (1987) 695-722

D2 ------ *$\mathcal{A}$-equivalence and the equivalence of sections of images and discriminants,* Singularity Theory and its Applications: Warwick 1989, Part I Springer Lecture Notes 1462 (1991), 93-121

D3 ------ *The versality discriminant and local topological equivalence of mappings,* Annales Inst. Fourier 40 (1990) 965-1004

D4 ------ *A Bezout Theorem for Determinantal Modules,* to appear Compositio Math.

D5 ------ a) *The unfolding and determinacy theorems for subgroups of $\mathcal{A}$ and $\mathcal{K}$,* Proc. Sym. Pure Math. 40 (1983) 233-254.

b) *The unfolding and determinacy theorems for subgroups of $\mathcal{A}$ and $\mathcal{K}$*, Memoirs A.M.S. 50, no. 306 (1984).

DM -- and Mond, D. *$\mathcal{A}$-codimension and the vanishing topology of discriminants*, Invent. Math. 106 (1991) 217-242

DG Damon, J. and Galligo, A. *A Topological Invariant for Stable Map Germs*, Invent. Math. 32 (1976) 103-132

DG2 ------ *On the Hilbert-Samuel Partition of Stable Map Germs*, Bull. Soc. Math. France 111, (1983) 327-358

Fa Falk, M. *A geometric duality for order complexes and hyperplane arrangements*, Europ. J. Combin. 13 (1992) 351-355

F Folkman, J. *The homology groups of a lattice*, J Math. and Mech. 15 (1966) 631-636

GW Gibson, C.G., et al. Topological stability of smooth mappings, Springer Lecture notes in Math. 552 (1976)

Gi Giusti, M. *Intersections complète quasi-homogènes: Calcule d'invariants*, Thesis Univ. Paris VII 1981 (also preprint Centre Math. de l'Ecole Polytechnique 1979)

GM Goresky, M. and Macpherson, R. Stratified Morse Theory, Ergebnisse der Math. und ihrer Grenzgebiete, Springer Verlag, Berlin, 1988

G Goryunov, V.V. *Singularities of Projections of Complete Intersections*, Jour. Sov. Math. 27 (1984) 2785-2811

Gr Greuel, G.M. *Der Gauss-Manin Zusammenhang isolierter Singularitäten von vollständigen Durchschnitten*, Math. Ann. 214 (1975) 235-266

GrH Greuel, G.M. and Hamm, H. *Invarianten Quasihomogener Vollständiger Durchschnitte*, Invent. Math. 49 (1978) 67-86

HLê Hamm, H. and Lê D.T. *Rectified Homotopical Depth and the Grothendieck Conjectures*, Volume in honor of A. Grothendieck, Birkhauser 1990

JS de Jong, T. and van Straten , D. *Dimension of the disentanglement component*, Singularity Theory and its Applications: Warwick 1989, Part I Springer Lecture Notes 1462 (1991), 199-211

Lê1 Lê D.T., *Le concept de singularité isolée de fonction analytique*, Adv. studies in Pure Math. 8, 1986, 215-227

Lê2 ---- *Calcul du nombre de cycles evanouissants d'une hypersurface complex*, Annales Inst. Fourier 23 (1973),261-270.

Lê3 ---- *Calculation of Milnor number of an isolated singularity of complete intersection*, Funct. Anal. and Appl. 8 (1974), 127-131.

LêT -- and Teissier, B. *Cycles évanescents, sections planes, et conditions de Whitney* II Proc. Sym. Pure Math. 44 Part II (1983) 65-103

LêT2 ------ *Limites d'espaces tangents en géométrie analytique*, Comm. Math. Helv. 63 (1988) 540-578

L Looijenga, E.J.N. Isolated singular points on complete intersections, Lecture notes in Maths. 77, London Math. Soc, 1984

L2 ---- *On the semi-universal deformation of a simple elliptic singularity: I. Unimodularity*, Topology 16(1977) 257-262.

MS Massey, D. and Siersma, D. *Deformation of Polar methods*, Annales Inst. Fourier 42 (1992) 737-778

Ma1 Mather, J.N. *Stratifications and mappings*, in *Dynamical Systems*, M. Peixoto (Ed.), Academic Press, New York, 1973, 195-232

Ma2 ------ *How to stratify mappings and jet spaces*, in Singularités d'Applications Differentiables, Plans-sur-Bex, Springer Lecture Notes 535, (1975), 128-176

Ma3 ------- *Notes on topological stability*, Harvard University, 1970.

Mi Milnor, J. Singular points on complex hypersurfaces, Ann. Math. Studies 61, Princeton, 1968

Mo1 Mond, D. *Vanishing cycles for analytic maps*, Singularity Theory and its Applications: Warwick 1989, Part I Springer Lecture Notes 1462 (1991), 221-234

Mo2 ------. *How good are real pictures?*, preprint 1992

N Nemethi, A. *The Milnor fiber and the zeta function of singularities of type $f = P(h, g)$*, Compositio Math. (1991) 63–97

OT Orlik, P. and Terao, H. Arrangements of Hyperplanes, Grundlehren der Math. Wiss. 300, Springer Verlag, 1992

OT2 ------- *Arrangements and Milnor Fibers*, to appear Math. Ann.

Sa Saito, K. *Theory of logarithmic differential forms and logarithmic vector fields*, J. Fac. Sci. Univ. Tokyo Sect. Math. 27 (1980) 265–291

Si Siersma, D. *Vanishing cycles and special fibres*, Singularity Theory and its Applications: Warwick 1989, Part I Springer Lecture Notes 1462 (1991), 292–301

Si2 ------ *Isolated line singularities*, Proc. Sym. Pure Math. 40 Part 2 (1983) 485- 496

Te Teissier, B. *Cycles évanescents, sections planes, et conditions de Whitney*, Singularités à Cargèse, Asterisque 7,8 (1973) 285–362

To1 Terao, H. *Arrangements of hyperplanes and their freeness I,II*, J. Fac. Sci. Univ. Tokyo Sect. Math. 27 (1980) 293–320

To2 ----- *The bifurcation set and logarithmic vector fields*, Math. Ann. 263 (1983) 313–321

To3 ----- *Generalized exponents of a free arrangements of hyperplanes and the Shephard–Todd–Brieskorn formula*,

Invent Math. 63 (1981) 159–179

Tg Tougeron, J.-Cl. Ideaux de Fonctions Differentiables, Ergebnisse der Math. Band 71 , Heidelberg-New York: Springer Verlag 1972

Wi Wirthmüller, K. Universell topologische triviale Deformationen, Thesis, Universität Regensburg

Z Zaslavsky, T. *Facing up to Arrangements: Face Count Formulas for Partitions of Space by Hyperplanes*, Memoirs. Amer. Math. Soc. 154 (1975)

Department of Mathematics
University of North Carolina
Chapel Hill, North Carolina 27599
U S A

Editorial Information

To be published in the *Memoirs*, a paper must be correct, new, nontrivial, and significant. Further, it must be well written and of interest to a substantial number of mathematicians. Piecemeal results, such as an inconclusive step toward an unproved major theorem or a minor variation on a known result, are in general not acceptable for publication. *Transactions* Editors shall solicit and encourage publication of worthy papers. Papers appearing in *Memoirs* are generally longer than those appearing in *Transactions* with which it shares an editorial committee.

As of May 31, 1996, the backlog for this journal was approximately 7 volumes. This estimate is the result of dividing the number of manuscripts for this journal in the Providence office that have not yet gone to the printer on the above date by the average number of monographs per volume over the previous twelve months, reduced by the number of issues published in four months (the time necessary for preparing an issue for the printer). (There are 6 volumes per year, each containing at least 4 numbers.)

A Copyright Transfer Agreement is required before a paper will be published in this journal. By submitting a paper to this journal, authors certify that the manuscript has not been submitted to nor is it under consideration for publication by another journal, conference proceedings, or similar publication.

Information for Authors and Editors

Memoirs are printed by photo-offset from camera copy fully prepared by the author. This means that the finished book will look exactly like the copy submitted.

The paper must contain a *descriptive title* and an *abstract* that summarizes the article in language suitable for workers in the general field (algebra, analysis, etc.). The *descriptive title* should be short, but informative; useless or vague phrases such as "some remarks about" or "concerning" should be avoided. The *abstract* should be at least one complete sentence, and at most 300 words. Included with the footnotes to the paper, there should be the 1991 *Mathematics Subject Classification* representing the primary and secondary subjects of the article. This may be followed by a list of *key words and phrases* describing the subject matter of the article and taken from it. A list of the numbers may be found in the annual index of *Mathematical Reviews*, published with the December issue starting in 1990, as well as from the electronic service e-MATH [**telnet e-MATH.ams.org** (or **telnet 130.44.1.100**). Login and password are **e-math**]. For journal abbreviations used in bibliographies, see the list of serials in the latest *Mathematical Reviews* annual index. When the manuscript is submitted, authors should supply the editor with electronic addresses if available. These will be printed after the postal address at the end of each article.

Electronically prepared papers. The AMS encourages submission of electronically prepared papers in $\mathcal{A}\mathcal{M}\mathcal{S}$-TeX or $\mathcal{A}\mathcal{M}\mathcal{S}$-LaTeX. The Society has prepared author packages for each AMS publication. Author packages include instructions for preparing electronic papers, the *AMS Author Handbook*, samples, and a style file that generates the particular design specifications of that publication series for both $\mathcal{A}\mathcal{M}\mathcal{S}$-TeX and $\mathcal{A}\mathcal{M}\mathcal{S}$-LaTeX.

Authors with FTP access may retrieve an author package from the Society's Internet node `e-MATH.ams.org` (130.44.1.100). For those without FTP access, the author package can be obtained free of charge by sending e-mail to pub@math.ams.org (Internet) or from the Publication Division, American Mathematical Society, P.O. Box 6248, Providence, RI 02940-6248. When requesting an author package, please specify AMS-TEX or AMS-LATEX, Macintosh or IBM (3.5) format, and the publication in which your paper will appear. Please be sure to include your complete mailing address.

Submission of electronic files. At the time of submission, the source file(s) should be sent to the Providence office (this includes any TEX source file, any graphics files, and the DVI or PostScript file).

Before sending the source file, be sure you have proofread your paper carefully. The files you send must be the EXACT files used to generate the proof copy that was accepted for publication. For all publications, authors are required to send a printed copy of their paper, which exactly matches the copy approved for publication, along with any graphics that will appear in the paper.

TEX files may be submitted by email, FTP, or on diskette. The DVI file(s) and PostScript files should be submitted only by FTP or on diskette unless they are encoded properly to submit through e-mail. (DVI files are binary and PostScript files tend to be very large.)

Files sent by electronic mail should be addressed to the Internet address pub-submit@math.ams.org. The subject line of the message should include the publication code to identify it as a Memoir. TEX source files, DVI files, and PostScript files can be transferred over the Internet by FTP to the Internet node e-math.ams.org (130.44.1.100).

Electronic graphics. Figures may be submitted tc the AMS in an electronic format. The AMS recommends that graphics created electronically be saved in Encapsulated PostScript (EPS) format. This includes graphics originated via a graphics application as well as scanned photographs or other computer-generated images.

If the graphics package used does not support EPS output, the graphics file should be saved in one of the standard graphics formats—such as TIFF, PICT, GIF, etc.—rather than in an application-dependent format. Graphics files submitted in an application-dependent format are not likely to be used. No matter what method was used to produce the graphic, it is necessary to provide a paper copy to the AMS.

Authors using graphics packages for the creation of electronic art should also avoid the use of any lines thinner than 0.5 points in width. Many graphics packages allow the user to specify a "hairline" for a very thin line. Hairlines often look acceptable when proofed on a typical laser printer. However, when produced on a high-resolution laser imagesetter, hairlines become nearly invisible and will be lost entirely in the final printing process.

Screens should be set to values between 15% and 85%. Screens which fall outside of this range are too light or too dark to print correctly.

Any inquiries concerning a paper that has been accepted for publication should be sent directly to the Editorial Department, American Mathematical Society, P. O. Box 6248, Providence, RI 02940-6248.

Editors

This journal is designed particularly for long research papers (and groups of cognate papers) in pure and applied mathematics. Papers intended for publication in the *Memoirs* should be addressed to one of the following editors:

Selected Titles in This Series

(Continued from the front of this publication)

556 **G. Nebe and W. Plesken,** Finite rational matrix groups, 1995

555 **Tomás Feder,** Stable networks and product graphs, 1995

554 **Mauro C. Beltrametti, Michael Schneider, and Andrew J. Sommese,** Some special properties of the adjunction theory for 3-folds in $\mathbb{P}^5$, 1995

553 **Carlos Andradas and Jesús M. Ruiz,** Algebraic and analytic geometry of fans, 1995

552 **C. Krattenthaler,** The major counting of nonintersecting lattice paths and generating functions for tableaux, 1995

551 **Christian Ballot,** Density of prime divisors of linear recurrences, 1995

550 **Huaxin Lin,** C^*-algebra extensions of $C(X)$, 1995

549 **Edwin Perkins,** On the martingale problem for interactive measure-valued branching diffusions, 1995

548 **I-Chiau Huang,** Pseudofunctors on modules with zero dimensional support, 1995

547 **Hongbing Su,** On the classification of C*-algebras of real rank zero: Inductive limits of matrix algebras over non-Hausdorff graphs, 1995

546 **Masakazu Nasu,** Textile systems for endomorphisms and automorphisms of the shift, 1995

545 **John L. Lewis and Margaret A. M. Murray,** The method of layer potentials for the heat equation on time-varying domains, 1995

544 **Hans-Otto Walther,** The 2-dimensional attractor of $x'(t) = -\mu x(t) + f(x(t-1))$, 1995

543 **J. P. C. Greenlees and J. P. May,** Generalized Tate cohomology, 1995

542 **Alouf Jirari,** Second-order Sturm-Liouville difference equations and orthogonal polynomials, 1995

541 **Peter Cholak,** Automorphisms of the lattice of recursively enumerable sets, 1995

540 **Vladimir Ya. Lin and Yehuda Pinchover,** Manifolds with group actions and elliptic operators, 1994

539 **Lynne M. Butler,** Subgroup lattices and symmetric functions, 1994

538 **P. D. T. A. Elliott,** On the correlation of multiplicative and the sum of additive arithmetic functions, 1994

537 **I. V. Evstigneev and P. E. Greenwood,** Markov fields over countable partially ordered sets: Extrema and splitting, 1994

536 **George A. Hagedorn,** Molecular propagation through electron energy level crossings, 1994

535 **A. L. Levin and D. S. Lubinsky,** Christoffel functions and orthogonal polynomials for exponential weights on [-1,1], 1994

534 **Svante Janson,** Orthogonal decompositions and functional limit theorems for random graph statistics, 1994

533 **Rainer Buckdahn,** Anticipative Girsanov transformations and Skorohod stochastic differential equations, 1994

532 **Hans Plesner Jakobsen,** The full set of unitarizable highest weight modules of basic classical Lie superalgebras, 1994

531 **Alessandro Figà-Talamanca and Tim Steger,** Harmonic analysis for anisotropic random walks on homogeneous trees, 1994

530 **Y. S. Han and E. T. Sawyer,** Littlewood-Paley theory on spaces of homogeneous type and the classical function spaces, 1994

529 **Eric M. Friedlander and Barry Mazur,** Filtrations on the homology of algebraic varieties, 1994

528 **J. F. Jardine,** Higher spinor classes, 1994

(See the AMS catalog for earlier titles)